Green Computing in Network Security

Green Engineering and Technology: Concepts and Applications

Series Editors: Brujo Kishore Mishra, GIET University, India and Raghvendra Kumar, LNCT College, India

Environment is an important issue these days for the whole world. Different strategies and technologies are used to save the environment. Technology is the application of knowledge to practical requirements. Green technologies encompass various aspects of technology which help us reduce the human impact on the environment and create ways of sustainable development. This book series will enlighten the green technology in different ways, aspects, and methods. This technology helps people to understand the use of different resources to fulfill needs and demands. Some points will be discussed as the combination of involuntary approaches, government incentives, and a comprehensive regulatory framework will encourage the diffusion of green technology, least developed countries and developing states of small island requires unique support and measure to promote the green technologies.

Convergence of Blockchain Technology and E-Business
Concepts, Applications, and Case Studies
Edited by D. Sumathi, T. Poongodi, Bansal Himani, Balamurugan Balusamy, and Firoz Khan K P

Handbook of Sustainable Development Through Green Engineering and Technology
Edited by Vikram Bali, Rajni Mohana, Ahmed Elngar, Sunil Kumar Chawla, and Gurpreet Singh

Integrating Deep Learning Algorithms to Overcome Challenges in Big Data Analytics
Edited by R. Sujatha, S. L. Aarthy, and R. Vettri Selvan

Big Data Analysis for Green Computing
Concepts and Applications
Edited by Rohit Sharma, Dilip Kumar Sharma, Dhowmya Bhatt, and Binh Thai Pham

Green Computing in Network Security
Energy Efficient Solutions for Business and Home
Edited by Deepak Kumar Sharma, Koyel Datta Gupta, and Rinky Dwivedi

Handbook of Green Computing and Blockchain Technologies
Edited by Kavita Saini and Manju Khari

For more information about this series, please visit: www.routledge.com/Green-Engineering-and-Technology-Concepts-and-Applications/book-series/CRCGETCA

Green Computing in Network Security

Energy Efficient Solutions for Business and Home

Edited by
Deepak Kumar Sharma, Koyel Datta Gupta,
and Rinky Dwivedi

CRC Press is an imprint of the
Taylor & Francis Group, an **informa** business

First edition published 2022
by CRC Press
6000 Broken Sound Parkway NW, Suite 300, Boca Raton, FL 33487-2742

and by CRC Press
2 Park Square, Milton Park, Abingdon, Oxon OX14 4RN

© 2022 selection and editorial matter, Deepak Kumar Sharma, Koyel Datta Gupta, and Rinky Dwivedi; individual chapters, the contributors

CRC Press is an imprint of Taylor & Francis Group, LLC

Reasonable efforts have been made to publish reliable data and information, but the author and publisher cannot assume responsibility for the validity of all materials or the consequences of their use. The authors and publishers have attempted to trace the copyright holders of all material reproduced in this publication and apologize to copyright holders if permission to publish in this form has not been obtained. If any copyright material has not been acknowledged please write and let us know so we may rectify in any future reprint.

Except as permitted under U.S. Copyright Law, no part of this book may be reprinted, reproduced, transmitted, or utilized in any form by any electronic, mechanical, or other means, now known or hereafter invented, including photocopying, microfilming, and recording, or in any information storage or retrieval system, without written permission from the publishers.

For permission to photocopy or use material electronically from this work, access www.copyright.com or contact the Copyright Clearance Center, Inc. (CCC), 222 Rosewood Drive, Danvers, MA 01923, 978-750-8400. For works that are not available on CCC please contact mpkbookspermissions@tandf.co.uk

Trademark notice: Product or corporate names may be trademarks or registered trademarks and are used only for identification and explanation without intent to infringe.

Library of Congress Cataloging-in-Publication Data
Names: Sharma, Deepak Kumar, editor. | Gupta, Koyel Datta, editor. |
Dwivedi, Rinky, editor.
Title: Green computing in network security: energy efficient
solutions for business and home / edited by Deepak Kumar Sharma,
Koyel Datta Gupta, Rinky Dwivedi.
Description: First edition. | Boca Raton: CRC Press, [2022] |
Series: Green engineering and technology: concepts and applications |
Includes bibliographical references and index.
Identifiers: LCCN 2021032244 (print) | LCCN 2021032245 (ebook) |
ISBN 9780367562922 (hbk) | ISBN 9780367562939 (pbk) |
ISBN 9781003097198 (ebk)
Subjects: LCSH: Computer networks–Security measures. |
Computer networks–Environmental aspects.
Classification: LCC TK5105.59.G69 2022 (print) |
LCC TK5105.59 (ebook) | DDC 005.8–dc23
LC record available at https://lccn.loc.gov/2021032244
LC ebook record available at https://lccn.loc.gov/2021032245

ISBN: 978-0-367-56292-2 (hbk)
ISBN: 978-0-367-56293-9 (pbk)
ISBN: 978-1-003-09719-8 (ebk)

DOI: 10.1201/9781003097198

Typeset in Times
by Newgen Publishing UK

Contents

Preface vii
Acknowledgments ix
Editor Biographies xi
List of Contributors xiii

Chapter 1 Green Computing at a Glance: A Solution for the
Next Generation 1

Rinky Dwivedi and Koyel Datta Gupta

Chapter 2 Empirical Study of Green Cloud Environment, Edge/Fog
Computing 9

Deepak Kumar Sharma, Rhythm Narula and Prahalad Singh

Chapter 3 Systematic Study of VANET: Applications, Challenges,
Threats, Attacks, Schemes and Issues in Research 33

*Amit Kumar Goyal, Gaurav Agarwal, Arun Kumar Tripathi
and Girish Sharma*

Chapter 4 Data Center Security 53

Shaik Rasool and Uma N Dulhare

Chapter 5 Energy-Efficient Network Intrusion Detection Systems in the
IOT Networks 79

Alka Singhal, Bhushan K Jindal and Veepsa Bhatia

Chapter 6 HomeTec: Energy Efficiency in Smart Home 95

*Ashish Sharma, Sandeep Tayal, Parth Rustagi, Priyanshu Sinha
and Rohit Sroa*

Chapter 7 Impact and Suitability of Reactive Routing Protocols,
Energy-Efficient and AI Techniques on QoS Parameters
of WANETs 105

Meena Rao and Richa Gupta

Chapter 8 Malicious Use of Machine Learning in Green ICT 119

Pragya Kuchhal and Ruchi Mittal

Chapter 9 Enhanced Framework for Energy Conservation and Overcoming Security Threats for Software-Defined Networks 141

Abhishek Kumar Gaur and Deepak Kumar Sharma

Chapter 10 Smart Shopping Trolleys for Secure and Decentralized Services 161

Jaspreet Singh, Charanjeet Singh, Yuvraj Singh, Chamandeep Singh and Monica Bhutani

Index 169

Preface

It is of immense pleasure to launch our book entitled *Green Computing in Network Security* that explores the idea of energy-efficient computing for network and data security. In this era, where digitization and communication of each professional or personal information have become a significant part of our lives, a stable and secured network is indispensable. Though there is no network fully immune from threats and attacks, a stable and efficient network security system is essential for shielding against data theft and sabotage. Such a system can secure shared data across the network. Network security consists of the strategies and practices adopted to avert and scrutinize any probable misuse and alteration of shared information or unauthorized access or even denial of network resources and services. Network security strategies like cryptography, use of anti-malware software and strong authentication means are important ways of achieving data privacy, integrity of data, authentication and non-repudiation. However, adopting automatization and networking has increased the emission of toxic carbon elements in the environment. To reduce the release of environment hazards and to improve the regulation of energy usage in data centers, green computing has become essential. This makes the book a pertinent choice for everyone from students to cyber security experts.

The book chapters have been contributed by scholars, researchers, academicians and engineering practitioners. This book focuses on the security threats involved in wireless networks, data centers, IoT and cloud computing and de-centralized services. The book includes the exploration of malicious use of machine learning in green Information and Communication Technology.

The book received plenteous abstract articles which were subjected to rigorous review procedures to ensure that the selected articles met the required quality standards.

We would extend our gratitude to everyone who has contributed directly or indirectly for the book. We express our sincere gratitude to all the authors and reviewers who have been committed towards shaping this book even after facing hardships due to the current pandemic situation. Our earnest thanks to the publisher CRC Press (A Taylor & Francis Company) for accepting our book proposal.

Editors

Acknowledgments

Our gratitude belongs to all the co-authors without whom these pages would be blank. We would also like to thank all the referees for their useful suggestions.

Editors

Editor Biographies

Deepak Kumar Sharma received the B.Tech in Computer Science and Engineering from Guru Gobind Singh Indraprastha University, M.E. in Computer Technology and Applications and Ph.D. in Computer Engineering from University of Delhi, India. He is presently working as Assistant Professor in the Division of Information Technology, Netaji Subhas Institute of Technology, Dwarka, New Delhi, India. His research interests include opportunistic networks, wireless ad hoc networks, sensor networks and network security.

Koyel Datta Gupta is a professional with an experience of 17+ years in academics. She has completed her B.Tech in Computer Science and Engineering from the University of Kalyani, West Bengal in 2003 and is a gold medalist in M.Tech (Computer Technology) from Jadavpur University, Kolkata (2007). She received her Doctorate in 2015 from Jamia Milia Islamia, New Delhi. Her areas of research include network security, digital signal processing, pattern recognition and machine learning. Dr. Koyel is currently working as Associate Professor in Maharaja Surajmal Institute of Technology (MSIT) (under the IP University), New Delhi. She is currently the Head of the Department of the Computer Science and Engineering of the institute. She has held various other positions in MSIT for the past 12+ years. She has published more than 25 research papers in reputed journals and conference proceedings and has also authored books.

Rinky Dwivedi has completed her B.Tech in Computer Science and Engineering from Guru Gobind Singh Indraprastha University, Delhi in 2004 and M.E. in Computer Technology and Application from Delhi College of Engineering, Delhi in 2008. She has received her Doctorate in 2016 from Delhi Technological University, New Delhi in the field of Agile Method Engineering. Her areas of research include software engineering and method engineering, agile software and machine learning. Dr. Rinky has over 14 years of experience in academics. She worked as Assistant Professor for 9+ years at Maharaja Surajmal Institute of Technology, Delhi. She has also worked as a full-time research scholar in

Delhi Technological University for two years and is currently working as Associate Professor in Maharaja Surajmal Institute of Technology, Delhi. She has published more than 20 research papers in reputed journals and conference proceedings and has also authored books.

List of Contributors

Gaurav Agarwal
Department of Computer Science and Engineering, Invertis University, Bareilly, India

Veepsa Bhatia
Southern Methodist University, TX

Monica Bhutani
Electronics and Communication Engineering, BVCOE, Delhi, INDIA

Uma N Dulhare
Muffakham Jah College of Engineering & Technology, Hyderabad, India

Rinky Dwivedi
Department of Computer Science and Engineering
Maharaja Surajmal Institute of Technology

Abhishek Kumar Gaur
Department of Information Technology
Netaji Subhas University of Technology

Amit Kumar Goyal
Department of Computer Applications, KIET Group of Institutions, Delhi-NCR, Ghaziabad, India.

Koyel Datta Gupta
Department of Computer Science and Engineering
Maharaja Surajmal Institute of Technology

Richa Gupta
Department of Electronics and Telecommunication Engineering
Maharaja Surajmal Institute of Technology

Pragya Kuchhal
Dept. of Information Technology
Netaji Subhas University of Technology, Delhi, India,

Ruchi Mittal
Dept. of Computer Science, Ganga Institute of Technology and Management, Jhajjar, Haryana, India.

Rhythm Narula
Department of Information Technology, Netaji Subhas University of Technology

Meena Rao
Department of Electronics and Telecommunication Engineering
Maharaja Surajmal Institute of Technology

Shaik Rasool
Muffakham Jah College of Engineering & Technology, Hyderabad, India

Parth Rustagi
Computer Science and Engineering, Maharaja Agrasen Institute of Technology, New Delhi, India

Ashish Sharma
Computer Science and Engineering, Maharaja Agrasen Institute of Technology, New Delhi, India

Deepak Kumar Sharma,
Department of Information Technology
Netaji Subhas University of Technology

Girish Sharma
Principal, BPIBS, Delhi, 110092, India

Priyanshu Sinha
Computer Science and Engineering, Maharaja Agrasen Institute of Technology, New Delhi, India

Chamandeep Singh
Electronics and Communication Engineering, BVCOE, Delhi, INDIA

Charanjeet Singh
Electronics and Communication Engineering, BVCOE, Delhi, India

Jaspreet Singh
Electronics and Communication Engineering, BVCOE, Delhi, India

Prahalad Singh
Department of Information Technology, Netaji Subhas University of Technology

Yuvraj Singh
Electronics and Communication Engineering, BVCOE, Delhi, INDIA

Alka Singhal
Jaypee Institute of Information and Technology

Rohit Sroa
Computer Science and Engineering, Maharaja Agrasen Institute of Technology, New Delhi, India

Sandeep Tayal
Computer Science and Engineering, Maharaja Agrasen Institute of Technology, New Delhi, India

Arun Kumar Tripathi
Department of Computer Applications, KIET Group of Institutions, Delhi-NCR, Ghaziabad, India

LIST OF REVIEWERS

1. **Nitin Gupta**
   Department of Computer Science and Engineering
   nitin@nith.ac.in
   NIT Hamirpur, HP, INDIA

2. **S. K. Dhurandher**
   Department of Information Technology, NSUT
   dhurandher@gmail.com

3. **Hardeo Kumar Thakur**
   CST, Manav Rachna University
   hkthakur@mru.edu.in

4. **Malaaya Datta Bohra**
   NIT, Silchar
   malayaduttaborah@cse.nits.ac.in

5. **Bhupender Kumar**
   KIET group of institutions, Ghaziabad
   bhoopendra.kumar@kiet.edu

1 Green Computing at a Glance
A Solution for the Next Generation

Rinky Dwivedi and Koyel Datta Gupta

CONTENTS

1.1 What Is Green Computing? Why Green Computing? How to Implement Green?...1
    1.1.1 Why Do Green?..2
    1.1.2 How to Implement Green?...2
1.2 Downsides of Digitalization and Its Impact..2
1.3 Nanotechnology as a Green Solution for Integrated Circuits................3
1.4 Energy and Climate Impact of Digital Technology................................4
    1.4.1 Direct and Indirect Impact of Digital Technologies..................4
        1.4.1.1 Carbon Emission at Data Centers..................................4
        1.4.1.2 Green Computing in Cloud Environment: Serverless Computers...4
1.5 Energy Efficiency at Homes..5
1.6 Malicious Use of Machine Learning in Green ICT.................................5
1.7 Conclusion..6
References..7

1.1 WHAT IS GREEN COMPUTING? WHY GREEN COMPUTING? HOW TO IMPLEMENT GREEN?

As per the IFG International Federation of Green ICT (Information and Communication Technology) and IFG standard, Green IT is the study and practice of environmentally sustainable computing of Information Technology.

Green IT "is the study and practice of designing, manufacturing, using and disposing of computers, servers and associated sub-systems—such as monitors, printers, storage devices and networking and communication system—efficiently and effectively with minimal or no impact on the environment" [1,2].

"Green Computing is the use of computing devices in an environmentally friendly way" [3].

DOI: 10.1201/9781003097198-1

1.1.1 Why Do Green?

To reduce e-waste, to prevent global warming, to combat climate changes, to better utilize energy and resources, to reduce operating and capital expenditures, to reduce the impact of hazardous materials on the environment.

1.1.2 How to Implement Green?

- *Using Energy Star labeled products*: Energy star labeled on appliances indicates its energy efficiency. It helps in reducing the overall energy consumption.
- *E-waste recycling*: E-waste recycling is the recycling of used electronic items; instead of throwing e-waste, it could be given to non-profit organizations and charities or submitted to the municipal or private recycling bodies.
- *Remote working*: Remote working is an arrangement where people can work from home using the internet, telephone and mail.
- *Cloud computing*: Whether done in a private or a public cloud configuration, cloud technology addresses two critical elements of green computing, i.e., energy efficiency and resource efficiency through "Virtualization" [4,5].

1.2 DOWNSIDES OF DIGITALIZATION AND ITS IMPACT

We are in a digital era where very sophisticated devices such as computers and mobile phones form a part of our daily life. These devices are simply amazing and have revolutionized the way humans work, make use of their free time and interact with each other. This transformation has been possible in just a few decades, thanks to the combined effort of scientists and engineers. However, as seen in any revolution in human history we should ask ourselves whether we were ready for all these changes and the negative consequences that should be identified and need to be fixed. This invokes various ideas, therefore, instances of security issues [6] associated with how electronic data handle, just think about the hacking scandal in the US elections philosophical, unethical issues seem so dominating by technologies might not be the best way to bring up next generation [7,8]. There is an additional drawback that has not been mentioned; yet that is the increasing concern with the associated total power consumption of computing devices and their implementation on the environment.

The IT revolution has made it possible to come up with solutions for the mitigation of integrated services. An integrated circuits are electronic circuits that forms part of computer; it is called integrated because the electronic components such as transistors are directly fabricated on the silicon chip. This is why they can be made in extremely small sizes with billions of them occupying a small area. Also these devices are light but powerful. However, a high processing power implies larger power consumption. Still, devices such as mobile phone consume less energy than many devices at home; however their number is increasing rapidly across the world because human's hunger for technology is currently endless to illustrate how important this problem is to look at big data centers. Data centers such as Google and Facebook receive large amount of information from web and social media. Thousands and thousands of computing units in data centers create a lot of heat, which is the reason why these data centers are normally kept in cold areas as they require a gigantic cooling infrastructure. In fact, according to some sources if we imagine that all data centers to be forming as a single country, then that country will be the fifth least energy consuming country in the world with these figures getting worse further. This is because of the vital importance to create new technologies that are greener and are based on completely different mechanisms.

1.3 NANOTECHNOLOGY AS A GREEN SOLUTION FOR INTEGRATED CIRCUITS

Many researchers worldwide are investigating different approaches like nanotechnology and magnetism. Nanotechnology is the area of technology that deals with the objects at nanoscale that is usually in the range between 1 and 100 nanometer. In order to have an idea what this scale really means, let's take the example of a 1 mm scale ruler. Now, imagine this scale into 1,000 times. One of this unit is micron which is the scale of bacteria and if we divide one micron into 1,000 times more, then that is a nanometer. So now it can be imagined how small it is. Silicon transistors forming part of the integrated circuits are already nanometric. Electric charge is used to store bits of information in such devices. In such memories, it is necessary to ensure continuous supply of energy to keep those bits in memory and not get erased over time. Moreover, every time a computation bit needs to be transferred from the memory to the processing unit and vice versa. This is how the computation has been performed. As discussed earlier, computer architecture can be more efficient if a continuous supply of energy is not needed to keep those bits in memory and CPU merged together. However for this to happen, a completely new paradigm to computation is required.

This is why magnetic nanostructures are considered [6] for green computation or green computing. Today, the world is focusing its attention on nanomagnetic wires. A magnetic wire is a standard wire having diameters in nanoscale. They are several microns long because of their very elongated shape; the multiple bits along its length move as tiny configured packages. When several small pulses of current are applied across the wires, they don't need a continuous supply of energy to keep those bits in memory. By modifying the shape and connecting the configurations, it becomes possible to perform computation through them.

Scientists have been investigating the properties of nanowires for more than a decade, discovering fascinating properties. These ICs must have higher storage capacities and for that an additional radical change is required in the form of a three-dimensional chip. This is again a challenging task because microelectronics companies so far are fabricating two-dimensional chips. So, adapting a whole new paradigm requires dynamic changes. The three-dimensional chips need to be manufactured at a massive scale to meet the requirements. This technology is facing several challenges like the material required to make the wires needs to be improved so that data flow is smoothened.

World is inching closer to have in our hand devices that are based on circuits that are greener and have novel functionalities. This will make possible that these devices are less dependent on energy resources, thereby becoming less of a concern for the environment of the planet.

1.4 ENERGY AND CLIMATE IMPACT OF DIGITAL TECHNOLOGY

If the entire loop of carbon emission is alienated into parts, three quarter of the emission comes from energy sectors and industries, while the remaining are from agriculture, forestry and land use and other waste products. Within the energy sector emission comes from various sectors that consume fossil fuels, industries, buildings, transportation, coal and gas power and fossil fuel production.

1.4.1 DIRECT AND INDIRECT IMPACT OF DIGITAL TECHNOLOGIES

Digital technologies have a very real carbon footprint from the energy used to power the data centers, data networks, computer to smart phones that we use every day [9,10]. These digital devices impact other sectors as well, for example, how GPS smart phone enables new transport services like UBER and OLA; this is an example of primary effect. However, these devices impact many sectors in secondary form as well.

1.4.1.1 Carbon Emission at Data Centers

Power is the largest expense in data centers. If proper measures are taken to reduce power consumption one can provide better cost-effective solutions [11]. It is estimated that 4.5% of global electricity consumption in 2025 will be attributed to data centers [12,13], thus emphasizing the need for looking for ways to curtail power consumption.

1.4.1.2 Green Computing in Cloud Environment: Serverless Computers

Serverless computers have opened a new paradigm in cloud computing thus enabling unmatched freedom and flexibility and drastically reducing the importance of always ON server architecture in certain application scenarios. This has increased the options available for deployment teams, especially for application scenarios that involve intermittent event-driven computing. Immediate benefits include infinite atomic scalability and high cost savings on resource consumption. One has to pay for actual consumption time and not total uptime.

Apart from infrastructure-as-a-service, platform-as-a-service and software-as-a-service, there is now a prominent class of function-as-a-service model based on serverless architecture. Serverless is also suited to modern API marketplaces. Now apart from the obvious benefits of cost reduction, inherent scalability, deployment speed, reduced code complexity and reduced time to market, serverless computing also has a tremendous positive impact on environment in terms of energy saving.

Energy saving is an attractive benefit and is a major interest in green computing initiatives.

In typical server settings, server remains powered up even though it may be idle for long duration. This has a huge impact on environment over the last few years with enterprises ending up being over-cautious and over-provision for serverless application scenarios. The mix of traditional server infrastructure, container orchestration and serverless function as a service will unleash new possibilities for larger and more complex applications. Serverless computing has certainly opened a path and realistically reduced power consumption at data centers.

1.5 ENERGY EFFICIENCY AT HOMES

Energy efficiency at homes not only reduces the carbon footprint but also improves the finances incurred [14]. There are several ways that, if implemented correctly, will reduce a great amount of carbon emissions at home. Some important measures are as follows:

- *Find better ways to heat and cool the house*: half of the energy used in homes goes to heating and cooling the environment inside. Energy can be saved by using ceiling fans in place of air conditioning everywhere. Periodically clean air filters set thermostats to appropriate temperature.
- *Install solar water heaters*: Install solar water heaters at homes. In cold and arid regions where solar energy is not enough, another possible solution is to install tankless water heaters. These heaters heat water directly without storage.
- *Replace incandescent lights:* The use of new technology such as light emitting diode (LED) and further advances in lighting control to reduce the time lights that are switched ON but not used can be implemented.
- *Proper insulation at home:* It is a most cost-effective way to make home comfortable and energy efficient. A tightly insulated home can improve air quality and reduce the utility bills.
- *Buy energy star appliances and install efficient showerheads and toilets:* Energy-efficient devices are approved by both Department of Energy and Environmental Protection Energy Agencies. These devices include everything from television sets to air conditioners, music systems and more. According to a survey [5], if only 10% start using energy-star appliances, it would reduce carbon emission of 1.7 million acres of trees.
- *Landscaping for energy efficiency:* Carefully positioning of trees and shrubs can save up to 25% of energy of household items. Energy-efficient landscaping has additional benefits such as lower maintenance and cost reduction in devices used for air cleaning inside and outside the home.

1.6 MALICIOUS USE OF MACHINE LEARNING IN GREEN ICT

The enormous rise in the use of web application by users for availing web-based services (banking services, e-commerce websites, etc.) and even for interacting with family and friends (Facebook, Instagram) has exposed their vulnerabilities to different web attacks. A lone web attack does not mean the effect of the attack will only be confided in the host machine; on the contrary, it can eventually spread in both internal and external networks, thereby compromising the security of the whole network. The attacker targeting the host system can gain the complete control of the infected system.

In the last few years, some of the prominent cyber-attacks include structured query language (SQL) injection, cross-site scripting, denial of services, man-in-the-middle and phishing. For years, cross-site scripting (XSS) has posed a major threat to web applications [15]. Open Web Application Security Project (OWASP), in its report (2015), states that SQL injection and XSS are the two most widespread attacks exploiting web application weaknesses to threats and exposing the clients' information.

In the current scenario the potential threat due to risks is increasingly putting organizations and users at a very high risk of heavy financial or valuable information loses. Presently, green classifier schemes are proposed for various attack prevention. Many experimental results also show that the proposed green classifier schemes can successfully prevent these attacks.

A lot of research has been carried out to detect web attacks using machine learning (ML) and deep learning techniques. ML-based algorithms attempt to detect the possible presence of any web attacks through exhaustive training.

Over the most recent couple of years with the attention on green computing, and less expensive process and capacity, there has been a flood of enthusiasm for ML in green ICT [1–3]. This is so because green ICT made the revolution in producing the modern energy-efficient processors, the density-to-performance ratio has also improved, the expense of saving and overseeing a lot of information has been significantly brought down, able to distribute compute processing across clusters of computers to analyze complex data in record time, and ML algorithms have been made accessible through open-source networks with enormous client bases [16]. However, there are constraints to use ML under the roof of green ICT. Lack of perception about green computing is perhaps the biggest obstruction in implementing green IT [17], Control on expanding necessities of heat evacuating device, which rises because of the rise in absolute energy utilization by IT devices and the ML schemes with high dataset.

Though it is seen that ML has revolutionized the world, it has many challenges such as resource constraints and majorly its malicious use by hackers who intentionally develop malicious techniques to hack data. Therefore, in the present scenario, different approaches in green ICT and ML need to be used abundantly that will help in reducing the hazardous causes to network and maximize energy efficiency.

1.7 CONCLUSION

In 2021, carbon emission is expected to fall by around 7% because of the pandemic crises, which will be the biggest annual drop in the history. However, the need is that carbon emission must fall at this pace in every coming year for the coming decades.

That means we need big structural changes in policies and infrastructure. Since the kind of behavior on economic changes that world has faced in 2020 are just sustainable and came at an incredible cost on human life and society. With governments spending trillions on recovery packages, an amount needs to be stipulated on clean energy as well to attain the required drop. However, if the authorities emphasize in bringing technologies and industries only, carbon emissions are likely to bounce back.

Green computing practices have become a necessity in reducing the carbon footprint. To achieve this, world should use today's technology responsibly in order that the coming generation can enjoy our new technological world.

REFERENCES

[1] A. Roy, A. Datta, J. Siddiquee, B. Poddar, B. Biswas, S. Saha, et al., Energy-efficient data centers and smart temperature control system with IoT sensing, 2016 7th Annual IEEE Information Technology, Electronics and Mobile Communication Conference (IEMCON), 2016, pp. 1–4.

[2] C. Peoples, G. Parr, S. McClean, B. Scotney, P. Morrow, Performance evaluation of green data centre management supporting sustainable growth of the internet of things, *Simulation Modelling Practice and Theory*, 34 (2013) 221–242.

[3] D. Gmach, Y. Chen, A. Shah, J. Rolia, C. Bash, T. Christian, et al., Profiling sustainability of data centers, 2010 IEEE International Symposium on Sustainable Systems and Technology (ISSST), 2010, pp. 1–6.

[4] A. Rahman, X. Liu, F. Kong, A survey on geographic load balancing based data center power management in the smart grid environment, *IEEE Communications Surveys & Tutorials*, 16 (1) (2014) 214–233.

[5] A. Jain, M. Mishra, S.K. Peddoju, N. Jain, Energy efficient computing-green cloud computing, 2013 IEEE International Conference on Energy Efficient Technologies for Sustainability (ICEETS), 2013, pp. 978–982.

[6] L. Lakhani, Green computing—a new trend in it, *International Journal of Scientific Research in Computer Science and Engineering*, 4 (3) (2016) 11–13.

[7] K. Nanath, R.R. Pillai, The influence of green IS practices on competitive advantage: mediation role of green innovation performance, *Information Systems Management* 34 (1) (2017) 3–19.

[8] C.A. Chan, A.F. Gygax, E. Wong, C.A. Leckie, A. Nirmalathas, D.C. Kilper, Methodologies for assessing the use-phase power consumption and greenhouse gas emissions of telecommunications network services, *Environmental Science & Technology*, 47 (2012) 485–492.

[9] Z. Alavikia, A. Ghasemi, Collision-aware resource access scheme for LTE-based machine-to-machine communications, *IEEE Transactions on Vehicular Technology*, (2018) 1–1.

[10] A. Singh, U. Sinha, D.K. Sharma, Cloud-based IoT architecture in green buildings, Advances in Civil and Industrial Engineering, IGI Global, 2020, pp. 164–183.

[11] D.K. Sharma, S. Agarwal, S. Pasrija, S. Kumar, ETSP: enhanced trust-based security protocol to handle blackhole attacks in opportunistic networks, Lecture Notes in Electrical Engineering, Springer, Singapore, 2019, pp. 327–340.

[12] R. Lu, X. Li, X. Liang, X. Shen, X. Lin, GRS: the green, reliability, and security of emerging machine to machine communications, *IEEE Communications Magazine*, 49 (2011).

[13] Á.M. Groba, P.J. Lobo, M. Chavarrías, Closed-loop system to guarantee battery lifetime for mobile video applications, *IEEE Transactions on Consumer Electronics*, 65 (1) (2019) 18–27.

[14] R. Petrolo, V. Loscrì, N. Mitton, Towards a smart city based on cloud of things, a survey on the smart city vision and paradigms, *Transactions on Emerging Telecommunications Technologies*, (2015).

[15] M.A. Hoque, M. Siekkinen, J.K. Nurminen, Energy efficient multimedia streaming to mobile devices—a survey, *IEEE Communications Surveys & Tutorials*, 16 (2014) 579–597.
[16] M. Dayarathna, Y. Wen, R. Fan, Data center energy consumption modeling: a survey, *IEEE Communications Surveys & Tutorials*, 18 (2016) 732–794.
[17] N. Kulkarni, S. Abhang, Green industrial automation based on IOT: a survey, *International Journal of Emerging Trends in Science and Technology*, 4 (2017) 5805–5810|.

2 Empirical Study of Green Cloud Environment, Edge/Fog Computing

Deepak Kumar Sharma, Rhythm Narula and Prahalad Singh

CONTENTS

- 2.1 Introduction 10
  - 2.1.1 Introduction to Cloud, Fog and Edge Computing 11
    - 2.1.1.1 Infrastructure as a Service (IaaS) 12
    - 2.1.1.2 Software as a Service (SaaS) 12
    - 2.1.1.3 Platform as a Service (PaaS) 12
  - 2.1.2 Fog Computing 13
  - 2.1.3 Edge Computing 13
  - 2.1.4 Relation Between Different Types of Computing 13
  - 2.1.5 Comparison 15
- 2.2 Principal Disciplines of Green Computing 17
  - 2.2.1 Introduction to Green Computing 17
  - 2.2.2 Necessity of Green Computing? 17
  - 2.2.3 Approaches to Implement Green Computing 18
    - 2.2.3.1 Improvements in Microprocessor 18
    - 2.2.3.2 Modifications in Cooling Systems 18
    - 2.2.3.3 Improvements in Network 19
    - 2.2.3.4 Improvements in Disk Storage 19
- 2.3 Attacks/Threats in Green Computing 19
  - 2.3.1 Types of Attacks on Cloud Computing 19
    - 2.3.1.1 Malware Injection Attacks on Cloud 20
    - 2.3.1.2 Denial of Service Attacks 20
    - 2.3.1.3 Abuse of Cloud Services 21
    - 2.3.1.4 Side-Channel Attacks 21
    - 2.3.1.5 Man-in-the-Middle Cloud Attack 21
    - 2.3.1.6 Wrapping Attacks 22
    - 2.3.1.7 Insider Attacks 22
    - 2.3.1.8 Service or Account Hijacking 22
    - 2.3.1.9 Advanced Persistent Threats (APTs) 22
    - 2.3.1.10 Specter Attack and Meltdown Attack 22

DOI: 10.1201/9781003097198-2

	2.3.2	Problems of Privacy and Security in Fog Computing 22
		2.3.2.1 Authentication/Verification 23
		2.3.2.2 Network Fortification 23
		2.3.2.3 Confidentiality or Privacy 23
		2.3.2.4 Access Control/Protected Data Storage 23
	2.3.3	Privacy and Attack Sensitivity of Edge Computing 23
2.4	Ensuring Security of Cloud/Fog-Based Solutions 23
	2.4.1	Security Measures 24
		2.4.1.1 Detect Intrusions 24
		2.4.1.2 Safe APIs and Access...24
		2.4.1.3 Implement Access Management 25
		2.4.1.4 Enhanced Security Frameworks...25
		2.4.1.5 Stronger Authentication 25
		2.4.1.6 Limiting Access to Cloud Services 25
2.5	Research Challenges in Green Computing 25
	2.5.1	Description of Challenges 26
		2.5.1.1 New Optimization Techniques...26
		2.5.1.2 New Efficiency Data Center..26
		2.5.1.3 Developing Green Maturity Model......................................26
		2.5.1.4 Information Resource Usage Optimization 26
		2.5.1.5 Network for Data Center Cooling..26
		2.5.1.6 Monitoring 26
	2.5.2	Green Computing Practices and Issues 27
2.6	Further Advancements...27
	2.6.1	Multi-Cloud Security Is Being Provided by Using Centralized Platforms by Organizations 27
	2.6.2	Precautionary Steps Are Being Taken by Organizations to Protect Data before It Arrives in the Cloud 27
	2.6.3	Organizations Have Increased Their Priority for Control Identity and Access Control in the Cloud...28
	2.6.4	SASE Is Gaining Popularity...28
	2.6.5	Combating Struggles of Cloud-Based Solutions................................28
	2.6.6	Avoiding SaaS 28
	2.6.7	Optimization of Internet 28
	2.6.8	Edge Is the New Cloud 28
2.7	Conclusion..29
References..29

2.1 INTRODUCTION

This section provides a brief introduction to cloud, edge and fog computing followed by their comparison.

2.1.1 Introduction to Cloud, Fog and Edge Computing

With the diffusion of computers around the 1990s, new ways were explored to provide users with large-scale computing power through time-sharing leading to the development of cloud computing. The objective of cloud computing is to permit clients to take advantage of the entirety of the innovations present, without the requirement for profound information about or ability with every single one of them. The cloud means to reduce expenses and help the clients center around their center business as opposed to being blocked by IT hindrances [1].

The characteristic of collecting data from multiple devices and sites allows an organization to enormously surpass the assets that would in some way or another be accessible to it, liberating them from the necessity to keep the framework on location [1,2].

Though cloud computing architecture consists of many components. It is broadly divided into two parts with both ends connected via a network like an internet [3].

1. Front end
2. Back end

Front end: It consists of applications and interfaces which are essential to access the cloud computing platforms that are the client part of the cloud computing system [3], for example, Google Docs and Gmail.

Back end: It consists of the cloud itself. It alludes to all the assets needed to lay down cloud computing services. It involves enormous information stockpiling, virtual machines, security systems and so on (Figure 2.1) [3].

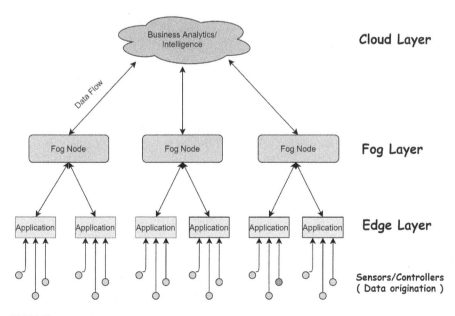

FIGURE 2.1 Cloud computing architecture.

Cloud computing includes both the software in the data centers that offer those types of assistance and applications conveyed as services over the web and the hardware.

The engineering of cloud registering is likewise classified by the following three sorts of conveyance models.

2.1.1.1 Infrastructure as a Service (IaaS)

Infrastructure as a Service consists of a single cloud layer where the cloud service provider resources are availed to contracted customers at compensation for each utilization [4]. This incredibly limits the requirement for immense starting investment in figuring computing hardware equipment like networking devices and servers.

2.1.1.2 Software as a Service (SaaS)

Software as a service (SaaS) consists of programming applications which are run on remotely located PCs managed by others [5]. A genuine illustration of such an application is Google Docs. SaaS offers a few key advantages, like immediate access and use of applications, openness from any machine that is associated, and furthermore that there is no presumably loss of information, as it is put away in the cloud [4].

2.1.1.3 Platform as a Service (PaaS)

PaaS is also a cloud-based environment that supports the building and deployment of cloud-based applications [5]. This service can be availed without purchasing any hardware, software, or even hosting. The primary advantage of using PaaS is the rapid deployment of applications, without stressing over any platform issues [5]. These service models abstract the hidden complexities and are much cheaper as well (Figure 2.2).

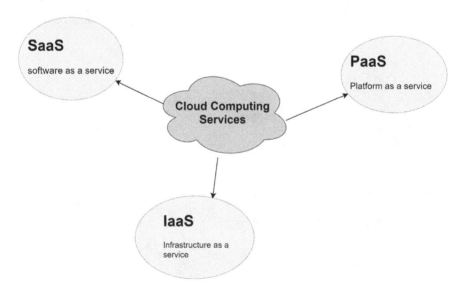

FIGURE 2.2 Types of cloud computing services.

The advent of cloud computing has drastically altered the conventional computing landscape by providing a level of abstraction or separation between the computing resources and the related technical framework, thus enabling demand-compatible self-service, location-independent resource management, ubiquitous network access, pay-per-use facility, rapid elasticity and requirement-dependent Quality of Service (QoS) [1].

Virtualization, one of the principal advancements in cloud computing, separates physical computing devices from many virtual devices making a versatile arrangement of different devices, allowing idle resources to be allocated easily and can likewise be utilized substantially more effectively [1].

Though cloud changed the whole computing landscape by providing a versatile and adaptable system for data analysis, it faced a variety of challenges as well. Security and communication challenges between local cloud and assets often lead to downtime. It was not apt for real-time automation which requires capturing data and analyzing it in real-time without any latency. Fog and edge computing were developed with the sole purpose to attenuate the risks.

2.1.2 Fog Computing

Fog computing is characterized by its decentralization of computing assets and locating them closer to the information generating sources. It is like an extension of cloud computing wherein we shift the processing of data closer to its source with the help of fog nodes [6,7]. Fog nodes play a critical function in universal working as they acquire facts from more than one source and examine that, thereby decreasing the number of facts dispatched to the cloud [8,9]. This facilitates in decreasing latency and thereby improving system response time, particularly in remote task-important packages. This also enhances the security of the business as by bringing data closer to the point of origination, they no longer have to send data through the comparatively insecure public internet (Figure 2.3).

2.1.3 Edge Computing

This computing methodology is one step ahead of fog computing with further advancements to make computing as close as possible to the user [7]. The primary advantage of edge computing is that it uses local places like the user's computer, IoT device, or edge server for computation rather than running them in the cloud minimizing the long-distance communication that used to happen between the user and the server [10]. Edge computing architecture (Data Processing Layer Stack) is represented in Figure 2.4.

2.1.4 Relation Between Different Types of Computing

Cloud, fog and edge computing techniques represent interdependent distinct layers of IT, where each one builds on the computing capacities of the previous layer [11]. Cloud computing with the integration of fog nodes leads to the formation of fog computing and further addition of edge leads to edge computing [12].

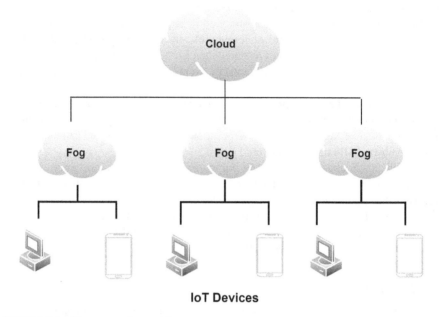

FIGURE 2.3 Fog computing architecture.

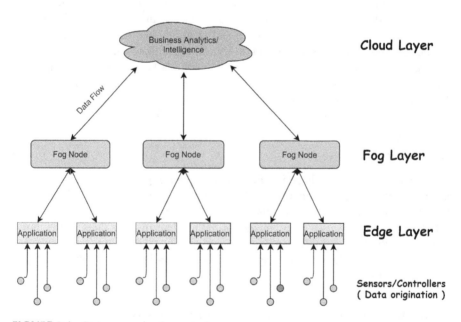

FIGURE 2.4 Data processing layer stack.

Empirical Study of Green Cloud Environment

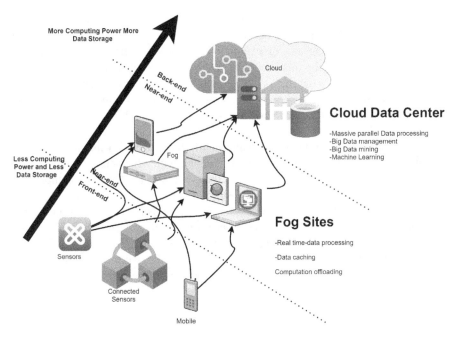

FIGURE 2.5 Relationship between cloud, fog and edge computing.

Despite the fact that these processing advancements contrast by their design and reason yet they regularly supplement one another. Both fog and edge computing offer comparable functionalities regarding pushing both insight and information close to the wellspring of the start of the information, be it vehicles, engines, speakers, screens, or sensors [13,14]. Both the advances influence the power of processing capacities to perform calculation errands that may have been done in the cloud without any problem. They assist organizations with decreasing their reliance on cloud-based stages for information handling and capacity, which frequently prompts latency issues, and can create information-driven choices quicker (Figure 2.5) [15].

2.1.5 COMPARISON

The essential distinction between different computing techniques is the location where processing of data takes place. Data processing takes place in the central cloud server for cloud computing while in edge and fog computing, processing takes place on the server/gadget and fog nodes, respectively [9].

With regard to the processing power and storage capabilities, cloud computing provides both much better and modern processing technological capabilities and huge storing capabilities compared to fog/edge computing in which major tasks take place on the devices or on the IoT sensor itself [16].

Latency, which tells us about the time taken by a data packet to traverse from starting node to ending node, is one of the major reasons for the development of edge/fog computing. Edge and fog being closer to the origin have far less latency as compared to cloud, thereby increasing the performance by many folds [17,18].

One more point of distinction is that though cloud computing requires consistent net access, at the same time rest two can work even without the internet making them more preferred instances where the IoT devices may not have consistent internet connection [19].

Talking about data protection, advancements have played a major role in this aspect. Fog/edge computing are much more secure with edge being the most secure as the data is stored in the device itself making it tough to change as compared to cloud computing which has a centralized structure [19].

Despite the fact that cloud processing is a promising registering viewpoint, which can give services to end clients concerning stage and programming, framework and supply applications with versatile assets successfully, however, it's anything but an adaptable course of action as there are still issues undetermined since IoT applications, all things considered, require portability, geo-appropriation and low slowness or latency (Figure 2.6) [14].

Features	Cloud Computing	Edge Computing	FogComputing
Network Connectivity	Wired or Wireless	Wireless	wireless
Access to service	Through server	Through Base station	At the fog computing centre
Agility	Slow	Fast	Fastest
Availability	Mostly Available	Mostly Available	Mostly Volatile
Bandwidth usage	High	Medium	Low
Computing Capacity	High	Medium	Low
Storage Capacity	High	Medium	Low
Connectivity	Internet	Follows Protocols	Follows Protocols
Distribution of Content	Edge devices	Distributed to base stations	Anywhere
Control	Centralized	Distributed till base stations	Distributed
Data analysis	Long term	Instant/Short term	Instant/Short term
Latency	High	No Latency	Low
Network traffic	High	Minimal	Low
Reliability	Low	High	Medium

FIGURE 2.6 Comparison table.

2.2 PRINCIPAL DISCIPLINES OF GREEN COMPUTING

2.2.1 INTRODUCTION TO GREEN COMPUTING

Green computing is characterized as the investigation and practice of planning, assembling, employing and disposing of PCs, networking devices and related subsystems efficiently and appropriately with negligible or no impact on the surroundings [20,21].

The key objectives of green computing include the augmentation of energy adequacy of item's lifetime, decreasing the utilization of unsafe materials, furthermore propelling the recyclability and biodegradability of outdated items [22]. In this section, we discuss possible designing systems and practices that make our use of computers energy efficient (Figure 2.7).

2.2.2 NECESSITY OF GREEN COMPUTING?

With the increase in demand of processing of large and at times distributed data sets that require real-time analysis and facilities like internet banking which requires security assurance, there has been an increase in the load on current data centers as these applications run for longer periods of time while these centers also host applications that require just serving of requests which run for a moment or two. These concerns have led to a great deal of computing specialist organizations like Google, Microsoft and IBM to set up data centers at numerous locations around the world [23][1].

With the aim of dealing with multiple applications simultaneously, organizations deal with challenges like on-request asset provisioning and assignment in time-evolving responsibilities by focusing only on the performance and compromising the power intake.

A massive data processing center requires a huge amount of energy, sufficient to power hundreds of homes. Organizations like Microsoft, Google, Amazon and plenty of different operators are spending large loads of cash on power payments [19]. Recently, as energy fees are increasing with dwindling availability, one needs to balance the optimization of data center resources while improving energy efficiency

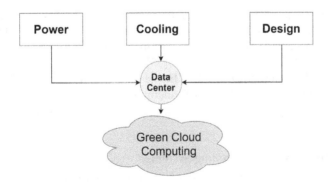

FIGURE 2.7 Green cloud computing.

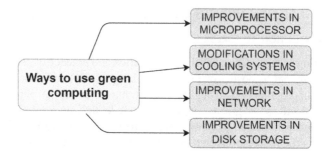

FIGURE 2.8 Ways to implement green computing.

Apart from the expensive maintenance cost, these data centers are threatening the environment. Being composed of numerous servers, they cause a tremendous carbon emission for cooling and controlling. These facilities now put more pressure on carbon emissions than Argentina and the Netherlands.

Bringing down the energy use and carbon footprint of data centers while meeting service is critical to make certain that the potential growth of cloud computing is imperishable. This approach caused the development of green computing.

2.2.3 Approaches to Implement Green Computing

From the previous section, we can conclude that tasks like cooling systems, computation processing, network and disk storage account for a majority of the power consumption (Figure 2.8) [24]. The accompanying subsection incorporates the new work done in these fields.

2.2.3.1 Improvements in Microprocessor

It is noticeable that microprocessors are the sole of the computational processing but with diminishing size of the transistors used in the microprocessors, there was a boom in electricity consumption via leakage currents to the factor that they fed on a greater energy than the real computational method itself [21,24]. It changed in the past due 2000s while the creation of recent substances has decreased the electricity dissipation of data centers [24]. Most outstandingly, the substitution of the Si_2 gate oxide, which is multiple nuclear layers thick with a definitely thicker layer of hafnium-based oxide leading to a considerable discount of the gate tunneling currents while preserving the overall electric efficiency of the transistor [25].

In modern-day microprocessor families, circuit architectures permit the power related to computational cycles and furthermore the spillage ability to be altered according to the requirements. These traits have significantly reduced energy consumption [26].

2.2.3.2 Modifications in Cooling Systems

Cooling systems devour almost 50% of the power consumption of data centers. Generally, the cooling framework of data centers is used to eliminate heat by constrained dissemination of a lot of chilled air [26]. By and by, researchers are attempting to utilize chilled-liquid cooling to cope with the huge influx of heat in densely packed servers. The designing of microchannel heat sinks is done in such a way that the thermal resistance between transistor and fluid is lowered to such an extent that even water with a temperature of 60–70°C can prevent the overheating of microprocessors. One of the major advantages of these hot water cooling systems is reduced dependence on chillers all year long reducing the energy utilization by up to half [24].

2.2.3.3 Improvements in Network

There has been a continuous increase in the cost of energy and network electrical components; correspondingly there has also been an increase in our susceptibility toward ecological issues, which has evoked curiosity in energy-efficient networking [27]. Earlier, network devices were equipped for apex load with the sole purpose of meeting high performance resulting in huge energy waste. Recently, there has been a shift toward using dynamically adaptive network equipment which adjusts according to the current traffic load and requirements, thereby reducing energy consumption as discussed above [25].

2.2.3.4 Improvements in Disk Storage

Storage is perhaps one of the highest energy consumers as compared to different components present in a data center, accounting for almost 27% of the total energy consumed [21]. Though there have been a lot of advancements in this field, one of the major breakthroughs appeared when the makers of solid state drives at Fusion-io were able to reduce operating costs and carbon footprint of MySpace data centers by 80% despite improving the performance speed [24]. They were able to achieve this by making use of smaller hard disks instead of heavy load-taking servers. Small hard disks consume less power per gigabyte due to the lack of any moving part with the additional benefit of having a lower carbon footprint [27].

2.3 ATTACKS/THREATS IN GREEN COMPUTING

Green computing is quickly turning out to be omnipresent while green figuring administrations are getting inescapable. There has been a rapid increase in cases related to attacks and threats against green services and computing. Organizations prefer data protection and they won't transfer their data to remote systems owned by cloud service providers if there are issues related to data security [28]. Data security, virtualization security, network security, identity management and application integrity are amongst the common security issues in green cloud computing which need to be addressed (Figure 2.9) [29,30].

This subject has a lot of challenges but the most common ones are listed below.

2.3.1 TYPES OF ATTACKS ON CLOUD COMPUTING

Cloud computing services are being attacked at an increasing rate; programmers are continually dealing with much more refined attacks [31]; furthermore, getting

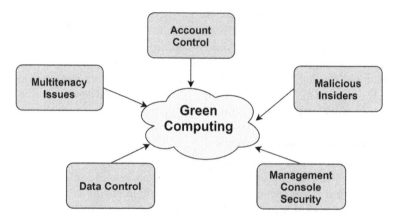

FIGURE 2.9 Major categorization of threats.

knowledge of common attacks assists the developers in designing much more secure outcome.

The major classifications of cyber-attacks on cloud computing are as follows (Figure 2.10).

2.3.1.1 Malware Injection Attacks on Cloud

These attacks on cloud computing are performed in order to gain control over the user's private data. Here, attackers insert a malicious job module to a PaaS or SaaS solution or a VM framework to an IaaS solution. If hackers are successful in hacking, the system redirects the client's private information to the hacker's module [31] and at times initiates the execution of malicious code.

Among malware injection attacks the most well-known are SQL injection attacks and cross-site scripting attacks. During a SQL injection, attackers target SQL servers with database applications. Sony's PlayStation web forum became the victim of a SQL injection attack in 2008 [28]. In cross-site scripting attacks, attackers add scripts (JavaScript, Flash) to a targeted web page. In 2011, researchers from Germany managed an XSS attack against the Amazon Web Services cloud computing framework [28].

2.3.1.2 Denial of Service Attacks

These sorts of attacks are intended to make a system less responsible and make administrations inaccessible to its users. These are specifically hazardous for cloud registering systems, as many users may suffer as a result of the overloading [31] of even a single cloud server. During overload, more computational power is provided by cloud services by including more superuser permissions and more VMs [28]. While attempting to forestall a cyberattack, the cloud system makes it seriously pulverizing. In the end, users are unable to access their cloud resources and the cloud system shuts down.

Empirical Study of Green Cloud Environment

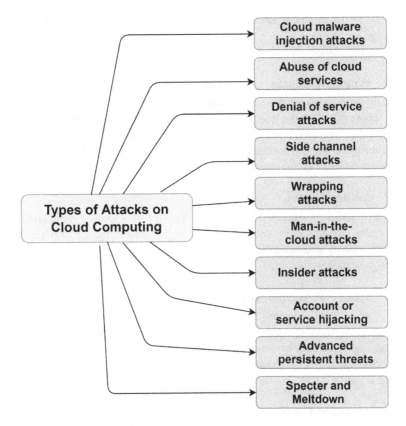

FIGURE 2.10 Types of cloud system attacks.

2.3.1.3 Abuse of Cloud Services

In abuse of cloud services, organizations, target users and even other cloud providers are being attacked by using cheap cloud services to make brute force and DoS attacks. Capacities of Amazon's cloud systems were exploited by two security experts using DoS back in 2010 [28]. They were able to make Amazon users unavailable on the internet at a very cheap cost of $6 to rent virtual services.

2.3.1.4 Side-Channel Attacks

Hackers can perform this when they are successful in placing malicious VMware on a similar host as an objective VM. In a side-channel attack, hackers target framework executions of cryptographic algorithms. Nonetheless, this sort of danger may be dodged with a more secure system implementation of cryptographic algorithms, which is covered in the latter part of this chapter.

2.3.1.5 Man-in-the-Middle Cloud Attack

Here, a synchronization token is implanted by the hackers in place of another token giving access to them. Dummy synchronization token can be placed making it impossible to detect any attack [28]. In fact, there are chances that hacked accounts may never be recovered. This attack takes advantage by misusing fragility in the cloud system.

2.3.1.6 Wrapping Attacks

Man-in-the-middle attack based on cloud systems is one of the types of wrapping attacks. These attacks exist due to a typical limitation of cloud systems of having an interface based on a web browser [31]. For instance, Amazon's cloud computing had a vulnerability in one of its interfaces which was detected back in 2009. This vulnerability allowed hackers to change messages to clients.

2.3.1.7 Insider Attacks

This sort of attack is executed by a cloud supplier or a worker associated with a cloud vendor having special privileges [31]. In this attack, the attacker deliberately disregards security strategies. To forestall pernicious action of this kind, cloud engineers design secure architecture.

2.3.1.8 Service or Account Hijacking

This is done by getting access to the client's credentials. From phishing to cookie poisoning as well as the use of spyware are part of the most used techniques. Hijackers can withdraw private information and compromise cloud services once it has been hacked. For example, in 2007 a Salesforce's employee got into a phishing scam.

2.3.1.9 Advanced Persistent Threats (APTs)

These threats let programmers constantly take private information put away in the cloud or use cloud administration in secret by important users. The span of these attacks permits programmers to shift to safety efforts against them. When denied access is established, hackers travel through server farm organizations and exploit network traffic.

2.3.1.10 Specter Attack and Meltdown Attack

Specter and meltdown cyber-attacks showed up recently and have just become another danger to cloud services. They take the assistance of malicious JavaScript code; hackers could peruse encoded information through data storage points using a semantic flaw in modern processors [28,31]. Both these two sorts of cyberattacks break the dependence among applications and the working framework, allowing hackers to pursue commands from the kernel. These attacks are causing a lot of stress to users because the latest security patches are not pre-installed among primitive users.

2.3.2 PROBLEMS OF PRIVACY AND SECURITY IN FOG COMPUTING

Calculation and computation power, finding information, and system accessibility closer to end hubs is being improved by an astonishing worldview of fog cloud computing which stretches cloud computing to networking edges [32,33]. But this system has some serious security issues listed below.

2.3.2.1 Authentication/Verification

Cloud service vendors commonly control data centers in cloud computing operations as well as deployment. Fog service vendors are unusual parties as a result of diverse deployment choices:

1. **Cloud solutions providers** generally need to spread their cloud solutions to the networking edges.
2. **Web service providers** generally have authority over home provisions or mobile data connections, which may put up with fog with their on-hand architecture.
3. **End-users** generally have close-by private cloud solutions. Thus they need to reduce the cost of proprietorship [32,34] and might need to transform their personal cloud solutions into fog and rent resources for additional profits.

2.3.2.2 Network Fortification

Fog computing has two primary drawbacks like wireless and remote framework security. Attacks like sniffing and jamming would in general be in the assessment scope of remote organizations [34]. For the most part, someone needs to follow the designs physically conveyed by a network supervisor and withdraw the network executive's traffic from general information traffic.

2.3.2.3 Confidentiality or Privacy

Leaking [34] of confidential or private information of users.

2.3.2.4 Access Control/Protected Data Storage

The fog has similar security perils because user data is generally outsourced and fog nodes are provided with complete control over user data [32]. Thus, it is hard to ensure data security because outsourced data can be used for malicious activities or can be lost.

2.3.3 PRIVACY AND ATTACK SENSITIVITY OF EDGE COMPUTING

Edge computing requires special encryption techniques independent of the cloud system in order to facilitate data transfer between various distributed nodes connected via the internet [35]. A shift from a centralized top-down model to a decentralized model is required because edge nodes are resource-constrained devices with limitations in security methods, whereas ownership of collected data can be shifted from end-users to service providers by keeping data on the edge [35].

2.4 ENSURING SECURITY OF CLOUD/FOG-BASED SOLUTIONS

Cloud service nature is very dynamic and it recesses the primitive security model used for on-site software. It must be known to users that cloud service providers cannot provide total security in their cloud system [36]. In fact, users are also responsible

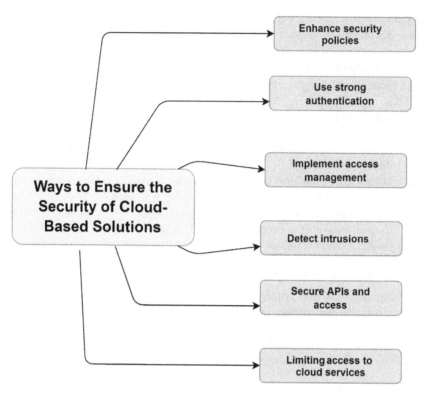

FIGURE 2.11 Means for securing cloud/fog solutions.

for some parts. The most secure way to safeguard user data is by implementing best practices to ensure the best possible security from the side of the cloud service provider (Figure 2.11) [37].

2.4.1 Security Measures

The following are the major ways by which cloud service providers ensure the security of their cloud system.

2.4.1.1 Detect Intrusions

This security measure is carried out by furnishing cloud system arrangement with a complete intrusion detection framework that can distinguish and can generate feedback of the wrong utilization of cloud administrations by hackers [36]. These frameworks also provide network checking ability.

2.4.1.2 Safe APIs and Access

When designing cloud servers, one ought to ensure that users can access the application via safe APIs only [38]. In order to accomplish this, we can limit the scope of IP addresses or provide access only via corporate networks.

2.4.1.3 Implement Access Management

Cloud developers should assign access credentials to various administrators so that users can only be assigned limited rights [36]. Besides, cloud administrators should empower favored clients so that they can set up the extent of clients' consents as indicated by their obligations inside the organization.

2.4.1.4 Enhanced Security Frameworks

Cloud developers limit the extent of their obligation regarding activities in the cloud security systems and guarding the client information [38]. Rather, they should make clients aware of their personal responsibilities toward cloud security and measures that they should adopt.

2.4.1.5 Stronger Authentication

Taking a password is the best widely recognized approach to get client's information and administrations in the cloud. Subsequently, cloud engineers should actualize solid confirmation and management of personnel identity. They must set up a multi-factor authentication system [38]. Different instruments that require both static passwords and captcha authentications must be used [36]. The latter affirm the client's accreditations by giving a one-time secret password or utilizing biometric plans or hardware tokens.

2.4.1.6 Limiting Access to Cloud Services

While developing cloud service frameworks, reduce event handler permissions to only those important for executing particular commands. Furthermore, restricting administration rights to only those cloud administrators who have been trusted to manage data security by the end-users is necessary (Figure 2.12).

2.5 RESEARCH CHALLENGES IN GREEN COMPUTING

A very large amount of energy resources for computing and air conditioning is required by data centers, supercomputers and real-time systems for high-performance cloud computing. From a green computing point of view, these are very serious issues and

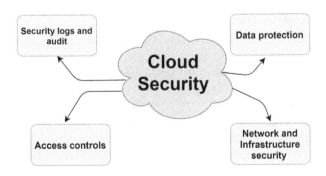

FIGURE 2.12 Types of green cloud security.

should be addressed very cautiously. Researchers are constantly working to reduce the effect of energy loss; yet there are domains to be explored.

2.5.1 Description of Challenges

Some prominent research domains in green computing are listed below [39].

2.5.1.1 New Optimization Techniques

Energy efficiency has become a technological issue of prime importance due to the rapid increase in computing environment and rising concern for energy conservation. In order to maximize net benefits, there must be a very delicate balance between energy, performance and temperature [13]. There is a very significant need to design innovative green models to save energy sustainably.

2.5.1.2 New Efficiency Data Center

Small data centers lack from bigger data centers in energy efficiency by a very big margin. There are various proposed scales to measure the efficiency of data centers. For example, PUE stands for power usage effectiveness [13] and is a measure of proportion consumption of the total power upon the net equipment power consumption.

2.5.1.3 Developing Green Maturity Model

With energy-saving index as the best measure of greenness, a complete equipment lifecycle is a prominent zone for green maturity models. There is a significant demand to propose innovative green maturity models for computing techniques and information technology-based organizations but improper research has led to their limitedness only in specific areas.

2.5.1.4 Information Resource Usage Optimization

The ability to analyze different database structures so that different databases could be interpreted irrespective of their data structures or storing mechanisms comes under this domain [13] so that data can be easily accessed and stored in any part of the world without repeated processing which causes energy loss.

2.5.1.5 Network for Data Center Cooling

Power utilization and conservation highly depend on the datacenter cooling system. Measuring the effectiveness and maintaining the baseline data center cooling is the major issue and must be addressed very cautiously.

2.5.1.6 Monitoring

Faster response to potential data center working threats is in high demand due to frequent cloud system failures. To help office administrators and IT staff control these issues, an ever-increasing number of administrators are going to DCIM's real-time, alarming/alert engine that offers permeability across all parts of the cloud system.

2.5.2 GREEN COMPUTING PRACTICES AND ISSUES

The implementation of green technologies at the organizational and industrial levels is emphasized by green computing practices. Global energy resources are continually declining and this is clearly evident from the fact that cost per unit of electricity charges is continuously increasing. Hence, both government sectors and private stakeholders must propose and practice remedial plans and strategies for green computing.

Practicing lightweight resource consumption protocols, disposal of electronic waste and resource-efficient cloud equipment are considered an integral part of green computing practices. Switching off cloud resources when not being utilized and scheduling cloud resources in power-saving mode are considered the best practices in green cloud computing.

Green computing also includes the management of e-waste produced by aging cloud resources. Further, older hardware parts generate greater energy loss and thus need disposal or replacements; also recycling is considered the most viable option. Green cloud computing has increased infrastructure and operational costs due to increased divergence from conventional computing techniques. Like a conventional grid, energy has a lower cost than renewable energy. In order to maximize comparable business incentives, efficiency of renewable energy source and storage techniques must be improved. There must be incentives to green cloud computing providers and users devised by public and governmental organizations.

2.6 FURTHER ADVANCEMENTS

Though the above discussed technologies have entered the industry, there is always a scope of development. In recent years, newer protocols have been proposed which make use of sophisticated technologies like Deep Learning [40,41] and Blockchain [42] for fostering security. Enhanced trust-based protocols are also being developed to handle various attacks. In the following section, we will be discussing further advancements in this field.

2.6.1 MULTI-CLOUD SECURITY IS BEING PROVIDED BY USING CENTRALIZED PLATFORMS BY ORGANIZATIONS

A unified way of securing data is being used by most service providers. This is being done by the centralization of compliance frameworks and security control [43]. For example, they are executing a cloud security access broker program that is present between cloud service clients and cloud applications, enforcing security frameworks and monitoring every process.

2.6.2 PRECAUTIONARY STEPS ARE BEING TAKEN BY ORGANIZATIONS TO PROTECT DATA BEFORE IT ARRIVES IN THE CLOUD

Arising mindfulness about information breaks and of new guidelines and norms pointed toward constraining associations to take care more for the information they rake in from customers has pushed organizations to begin ensuring data security before it arrives at the cloud [43].

2.6.3 Organizations Have Increased Their Priority for Control Identity and Access Control in the Cloud

A zero-trust model in which resources and access to data and services are restricted is being used by companies to improve their game on identity and access [43].

2.6.4 SASE Is Gaining Popularity

Secure and fast cloud adoption can be enabled using secure access service edge. This is done to ensure that data, applications and services can be accessed from anywhere and at any time by users or client applications.

2.6.5 Combating Struggles of Cloud-Based Solutions

Companies are highly attracted by the astonishing benefits of cloud computing-based security. Cloud computing is an intermix of reduced costs and economies of scale with better safety, improved security cognition [43] and effective compliance with industrial standards and governmental regulations. Instead of relying on primitive on-premises infrastructure, securing cloud systems requires new, innovative and different approaches [43]. Major software developers are constructing cloud-based security frameworks or trying to improve their existing SaaS techniques to secure their cloud-based systems.

2.6.6 Avoiding SaaS

SaaS is a highly established technique, but cloud data security issues are not addressed by them. This is properly addressed by IaaS and PaaS [43]. Organizations should try switching to IaaS and PaaS, step by step, in order to address major data security issues.

2.6.7 Optimization of Internet

The performance of internet websites is being enhanced by the utilization of fog computing by researchers in Cisco. Fog nodes can ease redirections, combining and executing content, fetching, stylesheet, scripts and images instead of going for a complete round trip for each HTTP request. Further, clients can be indexed on the basis of MAC addresses by fog nodes. Also, they can cache files, track user requests and determine local network strength.

2.6.8 Edge Is the New Cloud

Major product vendors like IBM, Intel, Dell and HPE [44] are enhancing their grip on edge computing with similar solutions like cloud equipment that can be placed anywhere and data center colocation (use of networking equipment and establishing private servers in a third-party data center) and content delivery networks [44]. Millions of local nodes are being used for edge computing services by major vendors.

2.7 CONCLUSION

One cannot deny the fact that cloud computing has drastically improved our work and life but it is contributing to environmental issues that nobody realizes.

Considering the above issues, researchers came up with green computing which corresponds to the efficient usage of available resources in a sustainable manner.

Though cloud computing is the most energy-efficient solution, it has a very serious drawback of limited security and low levels of encryption making organizations shift toward fog/edge computing. The potential of security mechanisms in the field of green computing is more than evident from the discussion throughout the chapter and subsequent developments have shown the acceptance of these mechanisms by the organizations. But this paradigm shift in the security mechanisms has led to major advancements in attacks as well. Extensive performance and security analysis show that the proposed security methods are resilient against failure and highly efficient, malicious data change attacks and even server overloading attacks but there is always room for improvement.

REFERENCES

[1] (Internet Source) Cloud computing https://en.wikipedia.org/wiki/Cloud_computing.
[2] D.K. Sharma, S. Agarwal, S. Pasrija, S. Kumar, "ETSP: enhanced trust-based security protocol to handle blackhole attacks in opportunistic networks," Lecture Notes in Electrical Engineering, Springer, Singapore, 2019, pp. 327–340.
[3] (Internet Source) Cloud computing architecture www.tutorialspoint.com/cloud_computing/cloud_computing_architecture.htm.
[4] (Internet Source) Three types of cloud computing service models www.bigcommerce.com/blog/saas-vs-paas-vs-iaas/#the-three-types-of-cloud-computing-service-models-explained.
[5] (Internet Source) Cloud computing infrastructure www.ibm.com/in-en/cloud/learn/iaas-paas-saas#:~:text=Infrastructure%20as%20a%20service%20(IaaS)%20is%20a%20cloud%20computing%20offering,within%20a%20service%20provider's%20infrastructure.
[6] J. A. S. K. C. M. Soumya Krishnan, "Green fog computing: a review on the basis of latency, energy, and e-waste," *IJAST*, 29 (3) (Mar) (2020) 5617–5625.
[7] The three layers of computing—cloud, fog and edge www.scc.com/insights/it-solutions/data-centre-modernisation/the-three-layers-of-computing-cloud-fog-and-edge/.
[8] S. Sarkar, S. Misra, "Theoretical modelling of fog computing: a green computing paradigm to support IoT applications," *IET Networks*, 5 (2) (Mar.) (2016) 23–29, doi:10.1049/iet-net.2015.0034.
[9] Differences between cloud, fog and edge computing in IoT—digiteum www.digiteum.com/cloud-fog-edge-computing-iot/.
[10] What is edge computing? www.theverge.com/circuitbreaker/2018/5/7/17327584/edge-computing-cloud-google-microsoft-apple-amazon.

[11] Cloud, fog, and edge computing: 3 differences that matter https://dzone.com/articles/cloud-vs-fog-vs-edge-computing-3-differences-that.
[12] Cloud, fog, and edge computing: 3 differences that matter www.networkworld.com/arCloud, Fog, and Edge Computing: 3 Differences That Matterticle/3224893/.
[13] G. Rahman, C.C. Wen, "Fog computing, applications, security and challenges, review," *IJET*, 7 (3) (Jul.) (2018) 1615, doi:10.14419/ijet.v7i3.12612.
[14] What is edge computing and why it matters www.networkworld.com/article/3224893/what-is-edge-computing-and-how-it-s-changing-the-network.html.
[15] What is edge computing and why it matters www.networkworld.com/article/3224893/what-is-edge-computing-and-how-it-s-changing-the-network.html.
[16] Cloud, fog, and edge computing: 3 differences that matter https://dzone.com/articles/cloud-vs-fog-vs-edge-computing-3-differences-that.
[17] M. Hassaan. "A comparative study between cloud energy consumption measuring simulators," *International Journal of Education and Management Engineering(IJEME)*, 10 (2) (2020) 20–27, doi:10.5815/ijeme.2020.02.03.
[18] Edge computing vs fog computing www.cmswire.com/information-management/edge-computing-vs-fog-computing-whats-the-difference/#:~:text=%E2%80%9CFog%20computing%20and%20edge%20computing%20are%20effectively%20the%20same%20thing.&text=So%2C%20with%20Fog%20computing%2C%20the,itself%20without%20being%20transferred%20anywhere.
[19] Cloud computing—fog computing & edge computing www.winsystems.com/cloud-fog-and-edge-computing-whats-the-difference/.
[20] Z. Ning, X. Kong, F. Xia, W. Hou, X. Wang, "Green and sustainable cloud of things: enabling collaborative edge computing," *IEEE Commun. Mag.*, 57 (1) (Jan.) (2019) 72–78, doi: 10.1109/mcom.2018.1700895.
[21] What is green computing? Scope, advantages & disadvantages https://bharatgogreen.com/green-computing/.
[22] A. Singh, U. Sinha, D.K. Sharma, "Cloud-based IoT architecture in green buildings," Advances in Civil and Industrial Engineering, IGI Global, 2020, pp. 164–183.
[23] N. Xiong, A. Vandenberg, W. Han, "Green cloud computing schemes based on networks: a survey," *IET Commun*, 6 (18) (2012) 3294–3300, doi: 10.1049/iet-com.2011.0293.
[24] Sustainable green computing: objectives and approaches www.ijarse.com/images/fullpdf/1510996834_244_IJARSE.pdf.
[25] Green computing: approaches and its implementations http://ijrar.com/upload_issue/ijrar_issue_1698.pdf.
[26] A.R.I. Mukaffi, R.S. Arief, W. Hendradjit, R. Romadhon, "Optimization of cooling system for data center case study: PAU ITB data center," *Procedia Engineering*, 170 (2017) 552–557, doi:10.1016/j.proeng.2017.03.088.
[27] (Internet Source) Green world: approaches to green computing http://greenworld2010.blogspot.com/2010/05/approaches-to-green-computing.html.
[28] M. Masdari, M. Jalali, "A survey and taxonomy of DoS attacks in cloud computing," *Security Comm. Networks*, 9 (16) (Jul.) (2016) 3724–3751, doi:10.1002/sec.1539.
[29] N. Karwasra, K., Mukesh, "Cloud Computing: Security Risks and its Future", *International Journal of Computer Science and Communication Engineering* IJCSCE Special issue on "Emerging Trends in Engineering" (2012).
[30] S.K. Sharma (New Delhi, India), B. Bhushan (Greater Noida, India), A. Khamparia (Punjab, India), P.N. Astya (Greater Noida, India), N.C. Debnath (Binh Duong Province, Vietnam) (Eds.), *Blockchain Technology for Data Privacy Management*, ch 5, CRC Press, Feb. 17, 2021, doi:10.1201/9781003133391.

[31] I. Butun, A. Sari, P. Österberg, "Hardware security of fog end-devices for the internet of things," *Sensors*, 20 (20) (Oct.) (2020) 5729, doi:10.3390/s20205729.

[32] S. Khan, S. Parkinson, Y. Qin, "Fog computing security: a review of current applications and security solutions," *J Cloud Comp*, 6 (1) (2017) 3–4, doi: 10.1186/s13677-017-0090-3.

[33] D.K. Sharma, A. Goel, P. Mangla, "Fog and edge driven security & privacy issues in IoT devices," in D. Gupta and A. Khamparia (Eds.), *Fog, Edge, and Pervasive Computing in Intelligent IoT Driven Applications*, Wiley, New Delhi, Dec. 07, 2020, pp. 389–408, doi:10.1002/9781119670087.ch21.

[34] J. Yakubu, S.M. Abdulhamid, H.A. Christopher, H. Chiroma, M. Abdullahi, "Security challenges in fog-computing environment: a systematic appraisal of current developments," *J Reliable Intell Environ*, 5 (4) (May) (2019) 209–233, doi:10.1007/s40860-019-00081-2.

[35] (Internet Source) Edge computing https://en.wikipedia.org/wiki/Edge_computing

[36] K. Jakimoski, "Security techniques for data protection in cloud computing," *IJGDC*, 9 (1) (2016) 49–56, doi: 10.14257/ijgdc.2016.9.1.05.

[37] The state of cloud security and privacy: 5 key trends to watch https://techbeacon.com/security/state-cloud-security-privacy-5-key-trends-watch.

[38] S. Khan, S. Parkinson, Y. Qin, "Fog computing security: a review of current applications and security solutions," *J Cloud Comp*, 6 (1) (Aug.) (2017) 4–10, doi:10.1186/s13677-017-0090-3.

[39] A. Riyal, G. Kumar, D.K. Sharma, "Background and research challenges for fog data analytics and IoT," *Studies in Big Data*, Springer, Singapore, 2020, pp. 313–340.

[40] M. Sharma, S. Pant, D.K. Sharma, K. Datta Gupta, V. Vashishth, A. Chhabra, "Enabling security for the industrial internet of things using deep learning, blockchain, and coalitions," *Trans Emerging Telecom Technol*, (Oct.) 1, 7–19 (2020), doi:10.1002/ett.4137.

[41] A. Dhiman, K. Gupta, D.K. Sharma, "Machine learning for fostering security in cyber-physical systems," *Security in Cyber-Physical Systems*, Springer International Publishing, New Delhi, 2021, pp. 91–122.

[42] B.K. Mishra (Gunupur, India), S.K. Kuanar (Gunupur, India), S.-L. Peng (Taipei, Taiwan), D.D. Dasig (Eds.), *Handbook of IoT and Blockchain*, CRC Press, Dasmariñas, Philippines, 2020.

[43] Bangui, H., Rakrak, S., Raghay, S., B., Barbora, Moving to the edge-cloud-of things: recent advances and future research directions, 7 (2018) 309, doi:10.3390/electronics7110309.

[44] H. Bangui, S. Rakrak, S. Raghay, B. Buhnova, "Moving to the edge-cloud-of-things: recent advances and future research directions," *Electronics*, 7 (11) (Nov.) (2018) 309, doi:10.3390/electronics7110309.

3 Systematic Study of VANET

Applications, Challenges, Threats, Attacks, Schemes and Issues in Research

Amit Kumar Goyal, Gaurav Agarwal,
Arun Kumar Tripathi and Girish Sharma

CONTENTS

3.1 Introduction 33
3.2 Application Areas 34
3.3 Layered Architecture of VANETs 36
    3.3.1 Components of VANETs 36
        3.3.1.1 Stationary Road Side Unit (RSU) 36
        3.3.1.2 On Board Unit (OBU) 36
        3.3.1.3 Application Unit (AU) 37
        3.3.1.4 Trusted Authority (TA) 37
    3.3.2 Transmission Methods 37
3.4 Challenges in VANETs 38
3.5 Issues in Security Requirements 40
3.6 Threats and Attacks 42
3.7 Emerging and Research Issues in VANETs 44
3.8 Research Work Based on Authentication in VANETs 44
3.9 Conclusion 47
References 50

3.1 INTRODUCTION

From the past few years, utilization of wireless technologies has increased the number of wireless products such as personal digital assistants, laptops and mobile phones. Also automobile companies have started using a widespread prospect for vehicles. Vehicular ad hoc networks (VANETs) [1], [2] have fascinated the automobile industry as well as research community. VANETs inherit the properties of MANETs [2], as shown in Figure 3.1, along with all the uncovered and revealed security mechanisms

DOI: 10.1201/9781003097198-3

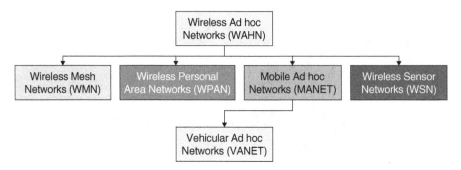

FIGURE 3.1 Classification of a wireless ad hoc network.

of MANETs. It provides various services related to vehicle safety, vehicle analytic, toll payment, congestion in traffic, enhanced navigation, services getting the location information related to the nearest gas/petrol station, lodge, or restaurant.

The traditional VANETs consist of mobile nodes, which are monitored by stationary sensors known as roadside units (RSU) [1]. Every vehicle is fitted with an on-board unit (OBU) [1], a wireless interface, which permits a vehicle to communicate with each other. A trusted third-party server known as authentication, authorization and accounting (AAA) [3] server is also placed along the roads. Communication in VANETs generally can be either: among vehicle to vehicle (V2V) [4] in which vehicles exchange information directly among themselves by sending and receiving the messages such as traffic related messages. Generally, this type of communication is a short-range vehicular communication, whereas the communication among vehicle to infrastructure (V2I) [4] allows to exchange information between vehicle and the RSUs. It may happen to share critical information such as road condition and speed limit. The remainder of this chapter is organized as follows: Section 3.2 expands application areas. Section 3.3 details about the layered architecture. Section 3.4 presents challenges that are there with VANETs. Section 3.5 describes the issues present in security requirements. Section 3.6 expands threats and attacks of VANETs. Section 3.7 highlights the emerging research areas; Section 3.8 describes the details of authentication security service and we conclude in Section 3.9.

3.2 APPLICATION AREAS

Several applications are possible with VANETs which can deliver a variety of information not only for drivers but also for passengers. These applications [5] allow users to exchange information related to traffic safety, traffic jam, upcoming toll, location detection, weather, parking and infotainment such as malls, theaters and restaurants. The applications of VANET largely fall into two categorical realms namely as safety-related applications and non-safety/convenience applications. Applications related to safety may reduce the probability of accidents which can cause death toll, whereas non-safety/convenience applications can improve traffic efficiency, passenger comfort by providing a variety of information such as climate information, congestion in traffic ahead, about the nearest parking available, petrol/gas stations, shopping malls, theater, hotels and restaurants. Table 3.1 categories various applications possible in VANETs.

TABLE 3.1
Application of VANETs

Application Type	Application	Description
Safety-related applications	Collision avoidance	Provide warning messages related to rules and regulations related to driving such as • Traffic signal violation • Stop signal violation • Intersection collision • Blind merge detection • Pedestrian crossing information • U turn/Left turn • Stop sign movement
	Public safety	Can provide critical information about • The emergency vehicle such as ambulance approaching • Anticipation of signal of emergency vehicle such as patrolling car • Post-crash signal which can save life of a person
	Sign extension	Can be used to provide various signs as warning messages which are very essential while driving • Seatbelt warning • In-Vehicle warning • High speed on curve • Low parking building • Low height bridge warning • Rash driving warning • Work/school zone warning
	Vehicle analytic and maintenance	Must assist driver about the vehicle by providing information about the vehicle itself such as • Service notification • Improper working of sensors • Safety recall warning • In-time overhaul warning
	Information of highways from other vehicles	Information from other vehicle which may be helpful to prevent accidents is necessary on highways such as • Blind spot analysis and warning • Vehicle changing lane • Cooperative accident warning • Warning of cooperative forward collision • Road condition information • Electronic light for emergency brake • Merge assistance for highways • Visibility enhancer in case of bad weather • Pre-crash detecting and warning • Vehicle by vehicle road condition announcement

(*continued*)

**TABLE 3.1 (Continued)
Application of VANETs**

Application Type	Application	Description
Convenience applications	Route diversion information	Information regarding a route is diverted for any purpose such as construction may save lot of time and effort
	Weather information	Bad weather information may be very critical and dangerous and if provided well in advance can help a driver from a traffic jam
	Traffic information	While roaming in urban areas the traffic condition if known in advance can both be a time and money saver
	Nearby restaurants/ petrol pump	On highways, the information of the nearest restaurants, gas/petrol pump is sometimes very crucial
	Electronic toll collection	Toll collection application can provide hassle free drive on highways
	Parking availability	Information about the nearest available parking plays a good impact while moving into the urban areas

3.3 LAYERED ARCHITECTURE OF VANETS

The architecture [2], [6], [7] of VANETs consists of a number of components, depicted in Figure 3.2, such as OBU, application unit (AU), RSU and trusted authority (TA).

3.3.1 Components of VANETs

3.3.1.1 Stationary Road Side Unit (RSU)

The RSUs are stationary computing devices, placed in specific locations after a fixed interval alongside the road. These units are used to communicate with the vehicle and can also be used to communicate with other networking devices. These devices use dedicated short-range communication (DSRC) [2] protocol established on the standard IEEE 802.11p for communication purpose.

3.3.1.2 On Board Unit (OBU)

Every vehicle is embedded with a tracking equipment, called OBU, that is based on GPS and primarily used for exchanging the information with other vehicles via their OBUs and/or stationary RSUs. Several sensors, providing inputs to OBU, include forward and backward sensors, sensor responsible for global positioning system (GPS), along with an event data recorder (EDR) [5], and a storage along with resource command processor (RCP) [5]. The OBU uses the IEEE 802.11p standard to connect with other OBUs and/or RSU.

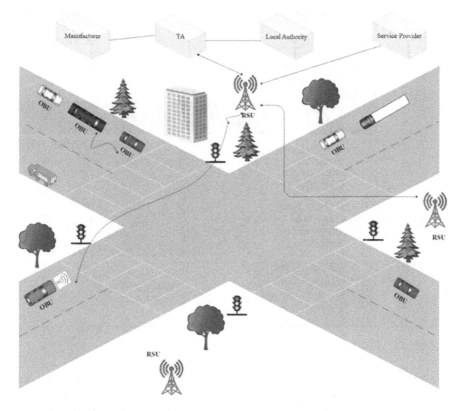

FIGURE 3.2 VANETs layered architecture.

3.3.1.3 Application Unit (AU)

The graphical interface acts as the application unit between the driver and OBUs which is used to retrieve the stored information such as the detailed information about the journey, speed information and information about traffic condition.

3.3.1.4 Trusted Authority (TA)

For managing the system, a central TA has the responsibility of authenticating the RSUs, OBUs and the vehicle. Thus, TA has the entire responsibility of securing the VANETs.

3.3.2 Transmission Methods

The VANETs use communication protocols [5]. Figure 3.3 represents the protocol such as DSRC protocol: IEEE 802.11p, and IEEE 1609.1-4, WiMAX: well suited for long-distance transmission. V2I, 3G: for flawless handoff, high-level latency, satellite: for universal, high cost, large propagation delay. DSRC, a dual-way short-to-middle-scale wireless communication tool, permits the safety and mobility of vehicles. In 1999, the band allocated from 5.850 to 5.925 GHz along with a spectrum of

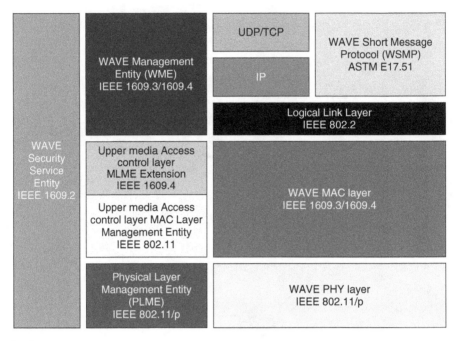

FIGURE 3.3 WAVE protocol.

75 MHz for DSRC by the US Federal Communication Commission (FCC) has seven channels of 10 MHz each in the 5.9 GHz band. Vehicle can use DSRC using transponders such as OBUs or RSUs. While in the V2V communication domain, vehicles exchange messages among themselves via OBUs to provide safety by alerting the driver about the possible danger, whereas in the V2I communication domain, an OBU communicates with stationary infrastructure known as the RSU for alerting the driver to safety risks such as an upcoming sharp curve or toll can be collected and/or be used for collecting parking payments. Wireless access for vehicular environment (WAVE) protocol: IEEE published the newest ITS standards, WAVE IEEE 1609, describing an architecture, the interface, the mechanism and a set of protocols (Figure 3.3). IEEE 802.11p: a new member 802.11p in IEEE 802.11 protocols by IEEE, for facilitating the vehicular communication network. WAVE short message service (WSMP) protocol: American Society for Testing and Materials (ASTM) developed E17.51 packets which specifies the priority, data transmission rate and capacity to travel long. WAVE management entity (WME) protocol: priority, the data transmission rate is registered by IEEE 1609.3 and IEEE 1609.4, WAVE security entity (WSE): encryption/decryption and key administration is handled by IEEE 1609.2.

3.4 CHALLENGES IN VANETS

The distinguishing characteristics of VANETs enact various challenges [8]–[10] for their implementation. These challenges range from technical that deals with the technical hindrances to social and economic as shown in Table 3.2.

TABLE 3.2
Challenges in VANETs

Challenge Type	Parameter	Description
Technical	High latency	The absence of the central coordinator in VANETs leads to the complex bandwidth management; consequently high-density area may be congested. Proper administration of bandwidth decreases the delay for the dissemination of transmitted messages
	Network heterogeneity	Geographical differences result in different infrastructures, security and privacy policies. Therefore, the protocols used by these heterogenous networks are different and leads to high latency
	High mobility	In VANETs, high-speed vehicles move on a predefined path resulting in the frequent change of network topology and channel condition. This may cause frequent disconnection and establishment resulting in higher latency as well as lot of effort required to authenticate vehicles moving on such a high speed
	Privacy	Privacy, an important aspect, means we should not disclose the location of the driver. But in case of casualty, the law enforcement agencies can decode the real identity and current position of the driver
	Computational capability	The most challenging issue is computational capability as number of sensors are placed in vehicles; high computational power can obtain the current status, speed and path
	Asymmetrical network concentration	The density of network varies as congestions, highways, narrow bridges, urban area and daytime are the factors that may cause higher density. Congestion and collision control are the most challenging tasks in VANETs
	Security	The most challenging issue is security which aims to provide services that incorporate services for authentication, availability of the network, data confidentiality, data integrity, nonrepudiation, privacy and anonymity of the driver, data verification, etc
	Routing protocol	Designing of efficient routing protocols is essential as it increases the reliability, robustness, scalability and decrease in latency
Social	Environmental impact	While deploying the VANETs, the impact on environment should be taken care of such as the electromagnetic waves used for communication have certain impact to the environment
	Tradeoff	Difficult for manufacturers to build up a network that delivers the traffic signal violation information because the driver rejects such type of monitoring. Conversely, the driver may be conscious of the warning message of police snare
Economic	ROI	For implementation of VANETs, lots of cost is involved, so customer and manufacturer both want their ROI

3.5 ISSUES IN SECURITY REQUIREMENTS

The security requirements in VANETs should be taken care of while designing an efficient network. Therefore, it is mandatory that the security requirements [11] [12], [13], described in Table 3.3, should be aligned with the overall operations of VANETs. Without taking care of these security requirements, a secure and efficient system cannot be implemented, which may lead to various threats or attacks.

TABLE 3.3
Issues in Security Requirements of VANETs

Issues/ Requirement	Description
Authentication	Authentication is the primary requirement for VANET's security, which guarantees that the messages should be forwarded by the genuine vehicle to reduce the attacks such as sybil, GPS Spoofing, Replaying, Tunneling, Message Tempering, and replication performed by the malicious vehicles. The identity of the vehicle should be verified first and distinguished the authenticated vehicles from unauthenticated one. It is essential because transmission of messages from unauthenticated malicious vehicles can cause severe penalties like human injuries, traffic interruptions and in radical cases which may even lead to death toll. Thus, authentication is a vital access control mechanism for deploying secure VANETs
Integrity	VANET is vulnerable to various active attacks such as masquerading, replay, message tempering and illusion, thus modifying the data being transmitted. Message integrity ensures that transmission of messages should not suffer from these types of attacks, i.e., messages cannot be altered while in transmission. Therefore, integrity is an indispensable requirement in vehicular communications
Accountability	Accountability, a crucial requirement, states that a node delivering the information should be obligated for its illegal actions. Malicious drivers need to be identified by the law enforcing agencies for their actions and therefore litigated for their unlawful activities. Once any mischievous activities are being performed by a malicious driver it cannot be denied
Non-repudiation	Once a legitimate sender has sent a message, non-repudiation ensures that it cannot refute that the message was not sent. Local authorities are capable to identify the originator of the message
Data consistency	It summarizes accuracy, authenticity, authorization and even integrity of data and states that all vehicles perceive a coherent viewpoint of the data
Message confidentiality	Sometimes it is required to communicate in private such as the law enforcement agencies need to find the adversary's vehicles, so they need to communicate with each other to communicate privately

TABLE 3.3 (Continued)
Issues in Security Requirements of VANETs

Issues/Requirement	Description
Privacy	Privacy is the major concern for the drivers as everybody wants that the past or future location of their vehicle should not be identified or predicted. Privacy ensures that the personal information such as real identity of the driver, personal statistics such as license plate number and location of the vehicle should not be leaked to the unauthorized people. However, to make drivers accountable for their actions and to trace them in case of casualty, the competent law implementation agencies should be able to track down the actual identity and/ or location
Entity authentication	It confirms that sender has started a fresh communication and gives an assurance to the recipient that the generated message is sent by a legitimate node of the network
Availabilty	Availability ensures that the attacks such as DoS, Jamming, Blackhole, Grayhole, Greedy behavior should not distress the functionality of the network and other applications which can be very disastrous in an emergency situation as the breakdown of instant reception of sent messages leaves the application of no use. Availability guarantees that instant information is available all the time to the legitimate user because any delay even if it is of fraction of second will make the message of no use
Access control	It is mandatory to establish proper access authorization control, which states that what each vehicle can do in and what types of messages can be generated by the specific vehicle, i.e., vehicles can function only according to the roles and privileges assigned during the authorization process
Scalabilty	Scalability allows the flexibility in the security architecture with an objective that the system design is significantly flexible enough so that expansion is possible
Storage constraints	Storage constraint requires that the storage space needed by a vehicle to store the private information such as identity of the vehicle, signature, group signatures, keying information should be minimized. This information is required by various cryptographic authentication techniques to provide secure communication
Robustness	System robustness confirms that even if intruders and faulty nodes are present in the network, the communication channel is secure enough
Unlinkabilty	Unlinkability which requires that messages and subjects should not be related to their actions. This is to prohibit for an adversary to deduce some information regarding the transmission that takes place between a sender and receiver
Minimum disclosure	For the sake of safety, the only minimum user information should be disclosed while transmission of information takes place

(*continued*)

TABLE 3.3 (Continued)
Issues in Security Requirements of VANETs

Issues/ Requirement	Description
Anonyamity	To maintain the driver's privacy, the actual identity must not be revealed; rather an anonymous identity mechanism should be adopted. However, for the accountability purpose the competent authorities track down the identity of faulty nodes to provide security to the vehicles
Conditional privacy	Conditional privacy states that though private records of the vehicle such as the vehicle's license plate number and location should be kept secret and should be prevented from an unauthorized access, in case of criminal action occurrence, a law enforcement agency has the full authority to divulge the identity of the vehicle
Liabilty detection	Though privacy is of atmost concern in VANETs accountability should be there for misconduct. For law enforcement agencies, it is required to identify vehicles to resolve the misconduct
Error detection	It ensures that malicious and specious transmission is detected by the system and such transmission will not take place at all
Efficiency	To make VANET systems efficient, we should be able to design a system that involves less time and low bandwidth. Therefore, the channel used for transmitting the messages will be having low delay
Real time assurances	It ensures that critical VANET applications such as safety-related communication should be able to broadcast the real-time information frequently to neighboring vehicles, i.e., the time kindliness of safety-related applications is there to avoid collision
Revocability	If a vehicle is found faulty for misconduct and or abnormal behavior, then that vehicle should be separated from the network by revoking its credentials and the network should be informed about this revocation of credentials

3.6 THREATS AND ATTACKS

Like any other wireless network, VANETs are vulnerable to numerous attacks [5], [7], [13], Figure 3.4 shows some of the attacks, which can affect the functioning and/or performance of VANETs. Sometimes, it is very difficult to identify the suspicious vehicles. To implement secure VANETs, it is necessary to determine who the attackers are, what types of attack are and what type of damage can be done by these attacks. In VANETs, the attackers are differentiated as follows:

  i. Depending upon the membership type of the node an attacker may be insider or outsider attackers. Insider attacker is a person who is an authenticated and authorized member of the network and having in-depth information of the network. On the other hand, outside attackers are unauthenticated persons having limited competence to attack as compared to insider attackers.

Systematic Study of VANET

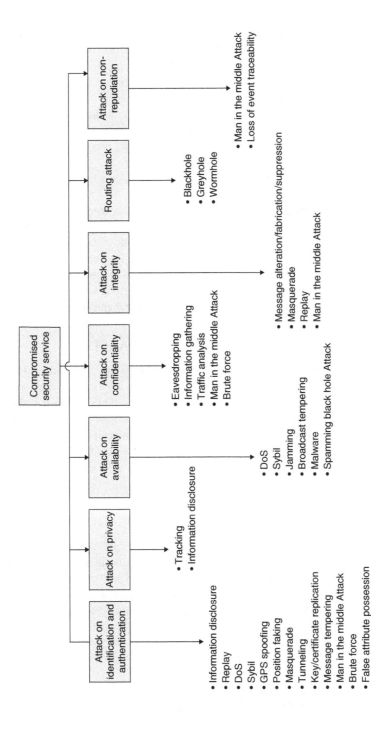

FIGURE 3.4 Different attacks and compromised services.

ii. Another classification is by the method of attack according to which attacker may be active or passive. In active attacks, the attacker generates in general bogus/counterfeit messages, or they obtain the packet destined to some node and either do not transmit it to the intended destination or transmit an altered message to the destination. On the other hand, the passive attackers only analyze the traffic and hardly participate in the actual communication taking place.
iii. The motivation behind the attack classifies attacker as the malicious attacker or rational attacker. The malicious attackers, having no intention of their personal aids, just extinguish the legitimate nodes present on the network, whereas the rational attackers, who are having personal yield associated with attack, attack the network unless until they gain access to the network and/or legitimate node.
iv. The scope of the attack decides whether an attacker may be a local attacker or extended attacker. In local attack, the attackers may gain access to limited resources only for some specific vehicles, whereas in extended type attack, the attackers may exploit full resources and even may control the entire network. To guard the network against different types of attacks, it is mandatory requirement that one must have knowledge of various security requirements issues and challenges in enforcing the security mechanisms.

3.7 EMERGING AND RESEARCH ISSUES IN VANETS

Due to high scalability nature of VANET where vehicles are being added in rapid manner, the most challenging tasks associated with such type of networks are security, reliability, performance, and Quality of Services (QoS). Over the past few years, many research opportunities [6], [7], [14] in the area of authentication, network security, routing protocol, QoS, performance-related parameters such as scalability, robustness, mobility model, fault tolerance, cooperative communication in VANETs, as depicted in Table 3.4, are being explored to acquire new designs and provide efficient services to the people.

3.8 RESEARCH WORK BASED ON AUTHENTICATION IN VANETS

To provide reliable and trustworthy vehicular communication, authentication is a vital step. It ensures that only legitimate vehicle can send the messages. Recently, several researchers have proposed various authentication schemes [23]–[31]. Some of these either provide authentication at node level, whereas some provide message-level authentication. Most of these algorithms use various cryptographic algorithms available. The broadcast messages in VANETs [32], [33] which are generally safety messages are vulnerable to threats of attacks. An attacker may alter a message generated by a legitimate vehicle, generate a bogus message, may contain changed location information and even of non-repudiation.

These safety messages require no confidentiality and/or privacy rather the main objective of these messages is that any how they should be delivered to every node present in the network without any delay and only authenticated vehicles should be

TABLE 3.4
Research Domain of VANETs

Key Area of Research	Description
Authentication	In order to provide secure and reliable vehicular environment, the authentication [15], [16], [17], [18], [19] service is essential, which ensures that only the legitimate users are allowed to exchange the messages. Recently, several researchers have proposed various authentication schemes. Some of these either provide authentication at node level, whereas some provide message-level authentication. The majority of these algorithm uses various cryptographic algorithm available
Network security	Providing an authenticated [3], [4], [5], [7], [9], [11], [12] secure network is key for the success of VANETs. As the vehicles in VANETs exchange information in an ad hoc manner, implementing an appropriate security protocol is must which can make the network secure
Routing protocols	An efficient and trusted routing algorithm can deliver the data packets to the appropriate destination well in time. Number of routing protocols [2], [20], [21] have already been proposed. Some protocols, used for MANETs, are topology based, i.e., proactive, reactive, or hybrid. Whereas some protocols use geographic and delay-tolerant approaches. Intense area of research is designing of an efficient routing algorithm that normalize the routing load so that it can route the data packets effectively and efficiently. Thus, the packets can reach to destination without any delay
Quality of services	QoS [22] is one of the most desirable activity in VANET. QoS guarantees low end-to-end delay, maximizing the average throughput, avoiding packet collision, minimizing the packet drop, and high connectivity time
Scalability and robustness	Designing and deploying a network, that is both scalable and robust [22], is undoubtedly an open area of research in VANET. Thus, resulting an operable and available network in extremely overloaded situations. It is always supposed that network must be fully functional in very low concentration regions such as rural areas and on highways, along with urban areas, where traffic density is very high
Fault tolerance	The very prominent research area in VANET is the designing a fault tolerant algorithm [22]. Like any other network, due to number of reasons such as hardware/software failure, tempering, a node in VANET is susceptible of failure at any time. Therefore, an effective recovery mechanism can only recover or protects the entire network from such types of failures

(continued)

**TABLE 3.4 (Continued)
Research Domain of VANETs**

Key Area of Research	Description
Cooperative communication	Establishing a communication among the vehicles that is cooperative enough is a key research area in VANETs. It decide the extent up to which a vehicle can exchange information with others. Previously proposed models of wireless technology may not be efficiently utilized for VANETs
Mobility model	Unlike conventional ad hoc networks, in which nodes are generally handheld devices having a limited mobility, the mobile nodes in VANET are extremely mobile. To develop a suitable and feasible model for VANET is an open research problem. A realistic mobility model, implementing a genuine traffic scenario, is the key for the success of VANETs. It can include different types of vehicles, roads maps, buildings, possible driving habits, varying traffic density, traffic rules and regulations

capable of sending them. The robust authentication mechanism can play a vital role to maintain the integrity of the message being in transmission as well. Authentication can be applied at the level of node known as node authentication or on message level called as message authentication. To detect and defend against an unauthorized node capable of sending the messages, numerous authentication schemes have been suggested by different researchers. Mainly cryptography techniques with some sort of signatures are generated at the sender side and later they are verified at the recipient side. Generally, these schemes are based on either signature-based schemes, cryptography-based schemes, or verification-based schemes. Figure 3.5 shows the authentication of the various schemes.

Signature-based authentication schemes further classified as certificateless signature scheme, ID-based signature scheme and authentication based on group signature scheme. Cryptography-based authentication schemes are using either asymmetric key cryptography that includes PKI certification based on elliptic curve digital signature algorithm (ECDSA), symmetric key cryptosystem that includes message authentication codes (MACs), hash functions, TESLA, TESLA++ and cryptography schemes based on IDs. Verification-based authentication schemes may use either batch verification or mutual verification for message authentication. Authentication schemes available must adhere to all the requirements and their performance can be measured based on various parameters such as the computation overhead measured in bytes and communication overhead measured in milliseconds, degree of authentication and ability of re-authentication and revocation in the case of malicious vehicles. Table 3.5 gives the insight of some of the research work carried out by various researchers.

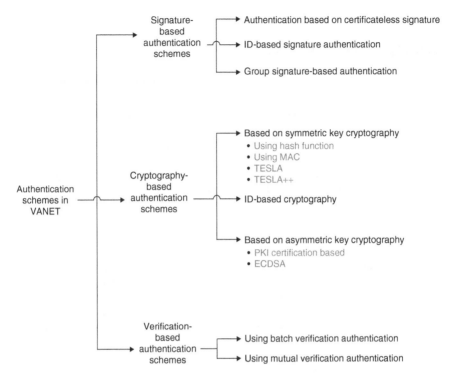

FIGURE 3.5 Various authentication schemes in VANETs.

3.9 CONCLUSION

Vehicular ad hoc networks (VANETs) consist of highly mobile vehicles that communicate with other vehicles and/or fixed infrastructure sporadically. The ad hoc nature of the network gives rise to risk of several attacks such as eavesdropping, impersonation, location detection, traffic analysis privacy and non-repudiation. These risks lead to various security vulnerabilities that may be for all the layers in the IP architecture stack. Deploying a secure network is the key for the success of VANETs. This chapter systemically reviews the applications areas, protocol stack associated with VANETs. Moreover, research opportunities in the implementation of Intelligent Transportation System are discussed. The implementation of VANET is one of the critical issues and suffers from a number of hindrances such as QoS, routing, security and latency. Furthermore, challenges associated with the implementation of VANETs are discussed. Authentication is the most critical issue associated with security. The chapter summarizes research work proposed by researchers related to various authentication schemes associated with VANET. At last, the chapter summarizes various authentication schemes proposed by researchers in VANETs. In future, we can incorporate 5G technology to serve the VANETs due to its silent features such as higher capacity, ultra-low end-to-end latency, higher data rate, massive device connectivity and quality of experience (QoE).

TABLE 3.5
Authentication Research work in VANETs

Research	Approach Used	Advantages/Disadvantages
[34]	Proposed a scheme based on group secret key agreement. To implement the anonymous authentication mechanism effectively and efficiently they allowed vehicles, RSUs and TA to communicate on scheme based on a batch authentication. All communicating entities shared a private key by a secure method	• The requirements related to communication and security are satisfied • Capable of protecting the privacy of vehicles • Able to trace actual identity when required
[35]	They have used a hybrid approach based on which a trustworthy cluster heads were elected. Which combined stability and trust factors. A adaptive trust function was derived for assessing the data trust among nodes. They introduced a timer for reducing the traffic control in the process of clustering and eliminated the competition involved in the nodes to be the cluster-head	• Improves the cluster stability effectively by guaranteeing improved vehicles cooperation and consistent sharing of data • To counter different types of attacks, the integration of other cryptographic parameters is not there
[36]	They proposed an authentication framework in which instead of using original IDs of vehicles, they used self-generated pseudo IDs in asymmetric cryptography. For authentication purpose, they proposed ID-based signature scheme and ID-based online/offline signature schemes. By using these schemes, they authenticated communication among all the entities such as vehicles, RSUs and TAs	• Impersonation attack and Sybil attack were resolved; they isolated the malicious nodes present in the network • Cooperative authentication among vehicles carried out without the intervention of RSUs • Location tracking capability to track attackers is not included • Identification of selfish nodes in the network is not there
[37]	They designed a lightweight authentication scheme based on time efficient stream along with bloom filters which is also a loss-tolerant authentication scheme. It prevents active attacks, has privacy-preserving feature and has better performance in comparison to the existing ones. They used the simulators such as NS-3 and simulation of urban mobility (SUMO) and evaluated the proposed model	• The proposed model tested for efficiency in terms of communication overhead for scenarios having both low and high densities • They found that the joint pseudonym achieved improved performance as compared without the pseudonym scheme • Outperforms the existing schemes based on parameters such as verification time and anonymity level

TABLE 3.5 (Continued)
Authentication Research work in VANETs

Research	Approach Used	Advantages/Disadvantages
[38]	They have proposed a framework in which they utilized the ID-based signature along with ID-based online/offline signature schemes. They used pseudonym-based schemes to ensure privacy preservation, and these pseudonyms are generated by public key cryptography	• Using PKC may cause more overheads • It is less efficient in case of a large-scale network
[39]	They proposed a scheme for symmetric operations based on message signing and verification. They proposed the strong privacy preservation and proposed a scheme which shows better performance with respect to delay, loss ratio and signing/verification of the messages can be achieved. They were not maintaining certificate revocation list (CRL). Thus, CRL overhead could be reduced	• They have reduced both computation and communication overhead. Along with this, overhead related to certificate management is also reduced • Used centralized key management center (KMC), which has disadvantages of bottleneck • The failure of KMC will result in total network failure
[40]	They estimated a round-trip distance that is verified by recording time	• Verification for dynamic network is missing • Use of digital signature may reveal identity of the node
[41]	They have used a fixed key infrastructure for finding impersonation attack and Sybil attack	• Inability to identify malicious node • If vehicles reach to different regions, finding of Sybil attack does not work accurately
[42]	They proposed a time efficient secure communication with privacy preservation scheme. Also, they have compared the proposed packet authentication scheme with existing public key-based cryptography	• The overhead that occurred in communication and computation can be significantly reduced in their scheme to verify each packet. Only a fast hash operation was needed • The revocation of safe key of compromised vehicles is a time-consuming process
[43]	They proposed an authentication scheme based on roadside unit (RSU). In the proposed scheme, for verifying the authenticity, RSUs are responsible. Along with this, RSUs are also responsible for alerting about the feedback to vehicles	• Revocation scheme that should revoke mischievous or faulty vehicles is missing • Fast disseminating revocation data is not addressed

REFERENCES

[1] H. Hartenstein, K.P. Laberteaux. 2008. A tutorial survey on vehicular ad hoc networks. *IEEE Communications Magazine* 46 (6) (June): 164–171.
[2] G. Karagiannis, O. Altintas, E. Ekici, G. Heijenk, B. Jarupan, K. Lin, T. Weil. 2011. Vehicular networking: a survey and tutorial on requirements, architectures, challenges, standards and solutions. *IEEE Communications Surveys and Tutorials* 13 (4) (July): 584–616.
[3] A.K. Tripathi, R. Radhakrishnan, J.S. Lather. 2017. Secure and optimized authentication scheme in proxy mobile IPv6 (SOAS-PMIPv6) to reduce handover latency. *International Journal of Computer Network and Information Security* 9 (10) (October): 1–12.
[4] Engoulou R.G., Bellaïche M., Pierre S., Quintero A. 2014. VANET security surveys. *Computer Communications* 44 (May): 1–13.
[5] M.S. Sheikh, J. Liang, W. Wang. 2019. A survey of security services, attacks, and applications for vehicular ad hoc networks (VANETs). *MDPI Sensors* 19 (16) (August): 3589–3628.
[6] W. Liang, Z. Li, H. Zhang, S. Wang, R. Bie. 2015. Vehicular ad hoc networks: architectures, research issues, methodologies, challenges, and trend. *International Journal of Distributed Sensor Networks* 11 (8) (August): 1–11.
[7] A.K. Goyal, G. Agarwal, A.K. Tripathi. 2019. Network architectures, challenges, security attacks, research domains and research methodologies in VANET: a survey. *International Journal of Computer Network and Information Security* 11 (10) (October): 37–44.
[8] M. Nidhal, J. Ben-Othman, M. Hamdi. 2014. Survey on VANET security challenges and possible cryptographic solutions. *Vehicular Communications* 1 (2) (April): 53–66.
[9] H. Hasrouny, A. Ellatif, C. Bassil, A. Laouiti. 2017. VANet security challenges and solutions: a survey. *Vehicular Communications* 7 (January): 7–20.
[10] S. Zeadally, R. Hunt, Y.S. Chen, A. Irwin, A. Hassan. 2012. Vehicular ad hoc networks (VANETS): status, results, and challenges. *Telecommunication Systems* 50: 217–241.
[11] F. Qu, Z. Wu, F. Wang, W. Cho. 2015. A security and privacy review of VANETs. *IEEE Transaction on Intelligent Transportation System* 16 (6) (December): 2985–2996.
[12] Z. Lu, G. Qu, Z. Liu. 2019. A survey on recent advances in vehicular network security, trust, and privacy. *IEEE Transaction on Intelligent Transportation System* 20 (2) (February): 760–776.
[13] S.S. Tangade, S.S. Manvi. 2014. A survey on attacks, security and trust management solutions in VANETs. Proceedings of the 4th IEEE International Conference on Computing, Communications and Networking Technologies, ICCCNT, (January), pp. 1–6.
[14] M. Sivasakthi, S. Suresh. 2013. Research on vehicular ad hoc networks (VANETs): an overview. *Journal of Applied Sciences and Engineering Research* 2 (1) (February): 23–27.
[15] G. Calandriello, P. Papadimitratos, J.P. Hubaux, A. Lioy. 2007. Efficient and robust pseudonymous authentication in VANET. Proceedings of the 4th ACM International Workshop on Vehicular Ad Hoc Networks, (September), pp. 19–28, https://people.kth.se/~papadim/publications/fulltext/-robust-pseudonymous-authentication-vanet.pdf.
[16] A. Studer, F. Bai, B. Bellur, A. Perrig. 2009. Flexible, extensible, and efficient VANET authentication. *Journal of Communication and Networks* 1 (6) (December): 574–588.
[17] R. Kalkundri, S.A. Kulkarni. 2014. A secure message authentication scheme for VANET using ECDSA. Proceedings of the 4th International Conference on Computing, Communications and Networking Technologies, ICCCNT, (January), pp. 1–6.
[18] A.K. Tripathi, S.K. Tripathi. 2018. A qualitative analysis of secured handover management schemes for mobile IPv6 enabled networks. International Conference on

Innovative Applications of Computational Intelligence on Power, Energy and Controls with Their Impact on Humanity, (November), pp. 1–8.
[19] A. Wasef, X. Shen. 2013. EMAP: expedite message authentication protocol for vehicular ad hoc networks. *IEEE Transactions on Mobile Computing* 12 (1) (January): 78–89.
[20] D.K.N. Venkatramana, S.B. Srikantaiah, J. Moodabidri. 2018. CISRP: connectivity-aware intersection-based shortest path routing protocol for VANETs in urban environments. *IET Networks* 7: 152–161.
[21] P. Fazio, F. De Rango, C. Sottile, A.F. Santamaria. 2013. Routing optimization in vehicular networks: a new approach based on multiobjective metrics and minimum spanning tree. *International Journal of Distributed Sensor Networks*, 2013: 1–13.
[22] M. Oche, A.B. Tambuwal, C. Chemebe, R. Md Noor, S. Distefano. 2018. VANETs QoS-based routing protocols based on multi-constrained ability to support ITS infotainment services. *Wireless Networks* 26 (October): 1685–1715.
[23] A. Studer, F. Bai, B. Bellur, A. Perrig. 2009. Flexible, extensible, and efficient VANET authentication. *Journal of Communication and Networks* 11 (6) (December): 574–588.
[24] J. Guo, J.P. Baugh, S. Wang. 2007. A group signature based secure and privacy-preserving vehicular communication framework. Proceedings of the Mobile Networking for Vehicular Environments (MOVE) Workshop in Conjunction with IEEE INFOCOM, 11 (May), pp. 103–108.
[25] N.V. Vighnesh, N. Kavita, R.U. Shalini, S. Sampalli. 2011. A novel sender authentication scheme based on hash chain for vehicular ad-hoc networks. Proceedings of the IEEE Symposium on Wireless Technology and Applications (ISWTA-2011), (September), pp. 25–28.
[26] S. Taeho, J. Jaeyoon, K. Hyunsung, L. Sung Woon. 2014. Enhanced MAC-based efficient message authentication scheme over VANET. Proceedings of the 7th International Multi-Conference on Engineering and Technological Innovation, IMETI, (July), pp. 110–113.
[27] P. Vijayakumar, V. Chang, L.J. Deborah, B. Balusamy, P.G. Shynu. 2016. Computationally efficient privacy preserving anonymous mutual and batch authentication schemes for vehicular ad hoc networks. *Future Generation Computer Systems* 78 (3) (January): 943–955.
[28] L. Lina, J. Castellà Roca, A. Vives Guasch, J. Hajny. 2012. Short-term linkable group signatures with categorized batch verification. Proceedings of the International Symposium Foundation and Practice of Security, (October), pp. 244–260.
[29] I. Ali, A. Hassan, F. Li. 2019. Authentication and privacy schemes for vehicular ad hoc networks (VANETs): a survey. *Vehicular Communications* 16 (April): 45–61.
[30] S.S. Manvi, M.S. Kakkasageri, D.G. Adiga. 2009. Message authentication in vehicular ad hoc networks: ECDSA based approach. Proceedings of the International Conference on Future Computer and Communication (ICFCC), (April), pp. 16–20.
[31] Y. Toor, P. Muhlethaler, A. Laouiti. 2008. Vehicle ad-hoc networks: applications and related technical issues. *IEEE Communications Surveys & Tutorials* 10 (3) (July): 74–88.
[32] S. Al sultan, M.M. Al doori, A.H. Al bayatti, H. Zedan. 2014. A comprehensive survey on vehicular ad hoc network. *Journal of Network and Computer Applications* 37 (January): 380–392.
[33] S.S. Manvi, S. Tangade. 2017. A survey on authentication schemes in VANETs for secured communication. *Vehicular Communications* 9 (July): 19–30.
[34] L. Lianhai, Y. Wang, J. Zhang, Q. Yang. 2019. A secure and efficient group key agreement scheme for VANET. *MDPI Sensors* 19 (3) (January): 1–14.

[35] S. Oubabas, R. Aoudjit, J.P. Joel, C. Rodrigues, S. Talbi. 2018. Secure and stable vehicular ad hoc network clustering algorithm based on hybrid mobility similarities and trust management scheme. *Vehicular Communications* 13 (July): 128–138.

[36] J. Jenefa, E.A. Mary Anita. 2018. Secure vehicular communication using ID-based signature scheme. *ACM, Wireless Personal Communications: An International Journal* 98 (1) (January): 1383–1411.

[37] S. Bao, W. Hathal, H. Cruickshank, Z. Sun, P. Asuquo, A. Lei. 2018. A lightweight authentication and privacy-preserving scheme for VANETs using TESLA and Bloom filters. *ScienceDirect ICT Express* 4 (4) (December): 221–227.

[38] J. Li, H. Lu, M. Guizani. 2015. ACPN: a novel authentication framework with conditional privacy preservation and non-repudiation for VANETs. *IEEE Transactions on Parallel and Distributed Systems* 26 (4) (April): 938–948.

[39] M. Wang, D. Liu, L. Zhu, Y. Xu, F. Wang. 2016. LESPP: lightweight and efficient strong privacy preserving authentication scheme for secure VANET communication. Springer-Verlag Wien 98 (7) (July): 685–708.

[40] C. Li, Z. Wang. 2012. Location-based security authentication mechanism for ad hoc network. National Conference on Information Technology and Computer Science, (November), pp. 382–385.

[41] M. Rahbari, M. Ali Jabreil Jamali. 2011. Efficient detection of sybil attack based on cryptography in Vanet. *International Journal of Network Security & Its Applications* 3 (6) (November): 185–195.

[42] Xiaodong L., X. Sun, X. Wang, C. Zhang, P. Ho, X. (Sherman) Shen. 2008. Timed efficient and secure vehicular communications with privacy preserving. *IEEE Transactions on Wireless Communications* 7 (12) (December): 4987–4998.

[43] C. Zhang, X. Lin, R. Lu, P. Ho, X. (Sherman) Shen. 2008. An efficient message authentication scheme for vehicular communications. *IEEE Transactions on Vehicular Technology* 57 (6) (November): 3357–3368.

4 Data Center Security

Shaik Rasool and Uma N Dulhare

CONTENTS

- 4.1 A Brief About Data Center 54
  - 4.1.1 Modern-day Data Centers 55
  - 4.1.2 Significance of Data Centers in Businesses 55
- 4.2 Data Center Design 56
  - 4.2.1 Objectives of Data Center Design 56
  - 4.2.2 Data Center Design Standards 57
    - 4.2.2.1 Power Needs of the Data Center 57
    - 4.2.2.2 Rack Volume 57
    - 4.2.2.3 Cooling 57
    - 4.2.2.4 Defects 57
    - 4.2.2.5 Integrated Data Center Protection and Maintenance 58
    - 4.2.2.6 Connection 58
  - 4.2.3 Design and Architecture of a Modern Data Center 58
- 4.3 Categorization of Data Centers 58
- 4.4 Data Center Structure 59
  - 4.4.1 Components of the Data Center 59
- 4.5 Data Center Management 60
  - 4.5.1 Data Center Management Problems 60
  - 4.5.2 Selecting the Best Data Center Infrastructure Management Solution 61
  - 4.5.3 Analyzing Data Center Infrastructure Management Software 61
  - 4.5.4 Uses of Data Center Infrastructure Organization Software 62
- 4.6 Data Center Standards 62
  - 4.6.1 Stand Tier of Uptime Institute 62
  - 4.6.2 ANSI/BICSI 002-2014 63
  - 4.6.3 ANSI/TIA 942-A 2014 63
  - 4.6.4 An International Standard: EN 50600 63
  - 4.6.5 Governing Standards 66
  - 4.6.6 Operating Standards 66
- 4.7 Green Data Center 66
  - 4.7.1 Significance of Green Data Centers 66
  - 4.7.2 Grading Environmental Sustainability 67
    - 4.7.2.1 Rapid Data Growth Thriving Power Demand 67
    - 4.7.2.2 Economic Scalability 67
    - 4.7.2.3 Guidelines for Selecting a Green Data Center 68

DOI: 10.1201/9781003097198-4

4.7.3 Green Data Center Benefits 68
    4.7.3.1 Further Testing of Biological Diversity 68
    4.7.3.2 Minimal Impact on the Environment 68
    4.7.3.3 Consumes Less Power 69
    4.7.3.4 Closing the Unused Servers 69
    4.7.3.5 Lower Financial Costs 69
4.8 Data Center Security 69
  4.8.1 Physical Protection 69
  4.8.2 Securing the Software 70
4.9 Critical Aspects for Securing Data Center 70
  4.9.1 Safety Information 70
  4.9.2 Monitoring and Authorization Access 70
  4.9.3 Improving Network Security 71
  4.9.4 Data Protection 71
  4.9.5 Infrastructure Redundancy 72
4.10 Data Center Security Standards 72
  4.10.1 Safety Level 72
  4.10.2 Logging in Access 72
  4.10.3 Video Surveillance 72
  4.10.4 Security Access Protocol 72
  4.10.5 Security Round the Day/Year 73
  4.10.6 RFID Asset Management 73
  4.10.7 Employee Verification 73
  4.10.8 Relieving Plans 73
  4.10.9 Multi-Factor Authentication 73
  4.10.10 Biometrics 73
4.11 Case Studies 74
  4.11.1 Microsoft Green under Water Data Centers 74
    4.11.1.1 Proof of Concept 74
    4.11.1.2 Power Wash and Data Collection 75
    4.11.1.3 Energy, Waste and Water 76
References 76

4.1 A BRIEF ABOUT DATA CENTER

Data centers worldwide are a reason for tremendous IT growth, extensive cloud infrastructure and a shift to virtualization [1]. Data center is a storage location for computer and communication equipment designed to record, store, process, send or access multiple files. Because of the high-quality supply that is usually stored online, data centers are at times referred to as service groups. These centers can store and assist online, provide e-mail and instant messaging, host cloud storage and applications, online services, competitive e-mail services, and create some other things which need to be explored. Almost every entity whether it is business or government requires own data center or wants someone else who may access it. Mostly, they build and maintain them indoors and some are widely used for hosting services such as Google, Sony and Microsoft [2] (Figure 4.1).

Data Center Security

FIGURE 4.1 Data center.

4.1.1 MODERN-DAY DATA CENTERS

Modern-day data center is very different to what it used to be years back. The organization has transitioned from traditional on-premise servers to virtual networks that boost applications and workloads across the entire physical structure and in multi-cloud environments. Today, data is available and can be accessed from multiple data centers, peripherals and communities, and in the personal cloud. Your data center needs to be ready to communicate from many places in the house and the cloud. Even various clouds can be stored in the data center. Once the applications are developed in the cloud, they use the data center services provided by the cloud service provider [3].

The data center is designed to build IT systems and other components such as:

- Power Distribution Unit (PDU)
- Routers
- Servers
- Uninterruptable power
- Cooling Units
- Storage

4.1.2 SIGNIFICANCE OF DATA CENTERS IN BUSINESSES

In the era of IT and industrial revolution, data centers are intended to support these business applications and activities:

- Productivity apps
- Big data, machine learning and artificial intelligence
- Customer relationship management
- E-mail support and file sharing options
- Enterprise resource planning and data centers
- Collaboration services, virtual desktops and communication apps

4.2 DATA CENTER DESIGN

Data center users today need more security for data center information. They are more focused on knowing if the data center is driven by a renewable source of energy. In many cases, they are looking at a more efficient server delivery, where one site could replace the older versions and focus more on air cooling and drought management systems. Many consumers share common and objective goals, and many combine external energy sources with toxic gases such as carbon monoxide. But people were not just buying change. The data center section changes internally and externally (Figure 4.2).

4.2.1 OBJECTIVES OF DATA CENTER DESIGN

Data center design objectives are known for improving performance, capability, flexibility, security and cost. We've all heard that before. Location is important. Having high performance and cost means choosing the location of the lowest cost and the

FIGURE 4.2 Data center infrastructure.

Data Center Security 57

most expensive equipment, as well as providing the best and most efficient (QoS) services available. That means you choose the right place for partners, businesses and clients.

But now virtualization and other new developments are changing the way we approach IT, forcing companies to rethink how design elements are placed. But today, technologies are changing the way we use IT, forcing businesses to rethink how we store data resources.

4.2.2 DATA CENTER DESIGN STANDARDS

4.2.2.1 Power Needs of the Data Center

Data centers are very hungry for power. Worldwide, storage consumes more than 400 terawatts of energy per year. However, just because data centers require a large amount of power does not mean that they have to do this. Buildings that efficiently distribute energy are carefully designed to ensure that they do not allow electricity to flow into heavy waste. They have also used modern technology to streamline energy regulation operations, as homes have become more and more robust. Many facilities also apply other design standards for green data centers to ensure that they promote sustainability and performance.

4.2.2.2 Rack Volume

The reason why data centers use so much energy is because companies have better data centers than before. These servers need higher energy, which entails that the data centers must have higher concentration of racks to adapt them. This change causes stress on data centers that do not have refrigeration spaces to fit high-performance devices. Large websites are often larger than their predecessors and force many organizations to reconsider whether they put valuable resources into the data layer. Unless the locker room is designed for today's stress-free shelf, it can be encouraged to use the right solution to reduce the limit and reduce workload.

4.2.2.3 Cooling

From innovations in the use of natural air leading to new themes, including outdoor air and water, air quality is a quality designed for the consideration of each data source. Although many companies still consider air-conditioned computer rooms (CRACs) to be relatively inadequate, the high demand for staff has now led to rapid development and adoption. New developments include direct cooling water and calibrated cold vector (CVC).

Although many facilities are unable to use these systems due to their high efficiency, they can still improve cooling performance with diagnostic and personalized tests. It is controlled by AI and technology. Recent Google research into device usage assures that this technology can reduce the cost overrun of key data.

4.2.2.4 Defects

Information received and crew work hours are important evaluations for any program resources from a data center. Every location should always have some kind of backup to protect customer information. For some of the base data centers, this means using a fault-breaking method that uses a series of backup operations to restore the device to its original state, and to avoid bugs. This service is only a millisecond. In some cases, software-based solutions can provide greater protection against a lack of tools to ensure that customers will always have access to the most important information [4].

4.2.2.5 Integrated Data Center Protection and Maintenance

Simple data generation should also be provided more protection to avoid data leak and can be achieved through integrated physical security. Regular legal checks are another important aspect of data security and consumer tools. Most of the shared clients address different management needs as part of their business; it is natural for data centers to lay the groundwork and operate with the following.

4.2.2.6 Connection

Crash storage is rarely used when there is a secure location for machine-only devices. Good websites set themselves apart by offering seamless connections and providing the public with a way to create the networks that can give the best insight into their industry. Operationally neutral data centers enable them to accurately build the solutions they own and to take benefit of cloud services while experiencing the same benefits of handling their private servers.

4.2.3 Design and Architecture of a Modern Data Center

The design of a modern data center starts with the precise installation of "good skeletons." The architecture of the data network should be extremely adaptable, as managers need to first see the potential to create adaptive raw spaces with swiftly evolving technology. The most successful people are those who anticipate trends (including AI, cloud computing, frontier computing and digital transformation) [5] (Figure 4.3).

4.3 CATEGORIZATION OF DATA CENTERS

There are various types of data centers and the costs involved vary widely. In a nutshell, this will provide simple, general instructions from the most traditional and expensive models to newer and cheaper ones.

1. **Business data center sites:** Developed, owned and controlled by companies, and designed for end users.
2. **Service provider:** This site information is controlled by a third party on behalf of the firm. The firm lends supplies and replacement equipment does not buy it.
3. **Colocation:** Within a colocation data center, companies list access to central data owned by others and located outside the home firm. The hosting center manages data management equipment: structure, cooling, bandwidth, security, etc., while the firm delivers and manages equipment, including servers, storage devices and rockets.

Data Center Security

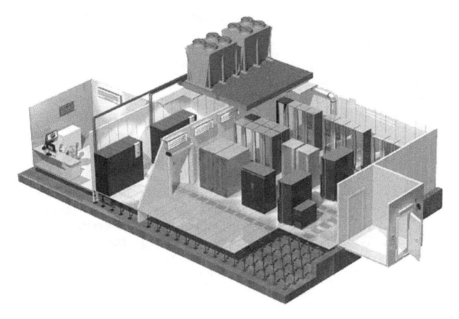

FIGURE 4.3 Modern data center concept design.

4. **Wholesale data center:** The reason it is popular in 2007 and 2008 is because some companies like Facebook, Microsoft and Yahoo are looking for more racks, but they want more space, and they always have no racks or machines. Create a new website. These centers are often referred to as MTDC data centers. MTDC leases usually charge 10,000 square feet of bonds from the government and science fiction companies, and the firm charges fees based on the funds provided to them.
5. **Cloud data center:** In this type of external data center, a cloud service provider like Microsoft Azure, Amazon Web Services or IBM Cloud, or another public cloud provider, retains data and applications.

4.4 DATA CENTER STRUCTURE

The data center structure includes all aspects of the main body of the data center. If the internal or physical structure are centrally located it forms data center infrastructure. The data center has two types of infrastructure: operational capacity and infrastructure designed to support the development of information technology. Computers and servers belong to the old group, such as cooling equipment, electricity and platforms. This is the end.

4.4.1 COMPONENTS OF THE DATA CENTER

Many data center sites are designed to withstand natural disasters such as floods, hurricanes, earthquakes, heat waves and so on. For this reason, many data warehouses are made of durable materials such as reinforced concrete [6]. The data center can cover one or more floors of a building. Some schools have high-quality facilities, which means that there is a gap between the actual quality and the quality that contains more information equipment. The space includes wiring, cabinet installation, refrigeration equipment and other equipment needed to maintain basic computer equipment. The advantage of enhancing the slide design is that it facilitates the use of connectors and security services.

1. **Data center hardware:** The main computer hardware in the common data center is the server, which is the center of the computer data center. Servers are usually stored on shelves and cabinets.
2. **Data center power:** Data centers contain large amounts of electricity in the form of power and other equipment. In addition to having enough electricity for the servers and wires to connect to the data center connected to the city electricity lot, the data center often includes power generation.
3. **Data center cooling:** To ensure that servers and other computing infrastructure are kept at an optimal temperature and are not overheated, most data centers have cooling equipment (including infrastructure) necessary to prevent the infrastructure from heating.
4. **Manage data centers:** Most data centers use a data management office or similar name; usually, there is a narrator in the area. Data center operations managers are responsible for maintaining the entire data center infrastructure. Journalists often use the centers to promote the essentials of their work.

4.5 DATA CENTER MANAGEMENT

4.5.1 DATA CENTER MANAGEMENT PROBLEMS

- Management of multiple devices, applications and sales activities related to many contracts, warranties, processes, paper licenses, repair and upgrade process.
- Experiment with multiple repositories with different plants, tools, machines and operations.
- Maintain the SLA in critical areas where data access, data storage, data rates and communications are available.
- Monitor changes due to lack or shortcomings of change management systems, obsolete equipment and inability to maintain infrastructure.
- Manage data center costs and budget cuts. In this case, resources are purchased at the "best cost"—whether they are distributed on the "side," electrical capacity or in one place. In addition, energy and air-conditioning costs are a percentage of investment.
- Accelerate implementation of new services and the use of lesser understanding of potential resources—strengths, stability and space, and unequal policies and procedures.

Data Center Security

4.5.2 Selecting the Best Data Center Infrastructure Management Solution

At present, the safety, high efficiency and green are becoming the key issues of system networking for data center infrastructure management [7]. Opting and applying the best tools for DCIM can increase ROI. As Gartner's researcher Chris Pettey stated in his report, "Other advances in energy efficacy are often accomplished via the use of DCIM software for a knowledgeable center". DCIM software delivers the essential connection between the operational requirements of equipment and physical facilities.

In the infrastructure management market, there are many types of free and open-source software solutions. Some considerations when choosing the best option for your business are listed in Table 4.1.

4.5.3 Analyzing Data Center Infrastructure Management Software

When evaluating new tools, do ask first if it can improve your business [8]. Questions for suppliers include:

TABLE 4.1
Issues to Consider When Selecting a Resolution

	Open-source	Closed-source
Cost	Tends to be free of charge, but be sure to consider the long term costs of in house technical and min	Provided for a cast by a vender
Service/Support	Can be limited or unstructured	Structured, professional service and support provided by a vendor
Upgrades/innovation	Innovation and new functionality may be shared by other users	Upgrades provided by vendor at regular intervals, but flexibility can be limited by to inability to change the source code
Usability	Depends on solution, but typically more focused toward technical developers rather than cannon users	Depends solution, but typically tested and designed by usability experts
Security	Depends on solution, but the software is not developed in a controlled, structured environment that focuses on security	Usually but with security in mind with security testing and audits
Customization	Access to source code allows deeper to customize	Usually provided at a cost
Risk	Risk usually assumed by user	Mitigated risk

- Can this tool be tailored to my everyday needs?
- Any adjustments can be made? Can I change the domain name according to my business? Each client has a different program that may or may not fit your case.
- Exactly how easily could it be read and applied? Research about usage and accessibility. Gartner's research first solved this problem, and there were some concerns when choosing DCIM software.
- Can commercial products connect future needs? Several resolutions provide a fair value and details or tools to make it happen.
- What is the guarantee or training in data center management provided by the provider?
- Can you customize your software to include old, new and limited content?
- What are the plan and the right time to establish the right income?

4.5.4 Uses of Data Center Infrastructure Organization Software

The functions and elements of DCIM emulsions vary, but the power consumption and regulatory control can reduce costs. Using DCIM is also very profitable for your business because it gives you the following:

- Uncover cost saving prospects
- Lower risk
- Evaluate transformation impact
- Enhance ease of use
- Enhance performance
- Systematize and enhance capacity planning and projecting
- Support IT business goals
- Lower power utilization
- Enhance asset consumption
- Supervise and enhance availability
- Reduction of equipment supply/discarding obstacles
- Upgrade overall IT service delivery

4.6 DATA CENTER STANDARDS

It is best practice to make sure you do everything in your power to do so. The best way is to say different things to different people and organizations. Compliance with applicable laws and standards is a clear guide to designing or updating a data sheet. The classification of data centers and assets may be based on national regulations (required) such as NFPA regulations, and local regulations such as the National Energy Conservation Association Code.

4.6.1 Stand Tier of Uptime Institute

This International Standard develops a performance-focused approach to information center operations in planning, construction and decentralization activities in four phases. Paragraph/reliability: The layers in the table below are relative, and you can

Data Center Security

find a better definition in the white letter. The roots of the "Uptime Foundation" as a group of users of information centers have grown since the first group survived and compared the reliability of information centers. It is a profitable entity that will certify facilities in accordance with its standards to ensure that the quality is generally criticized.

Tier Rating	Tier 1	Tier 2	Tier 3	Tier 4
Active Capacity Components	N	N+1	N+1	N after Failure
Distribution Paths	1	1	I. Active + 1 Alternate	2 Active
Concurrently Maintainable	No	No	Yes	Yes
Fault-tolerant	No	No	No	Yes
Compartmentalization	No	No	No	Yes

4.6.2 ANSI/BICSI 002-2014

Quality assurance locations and quality data implementation: This framework contains key elements of the layout, production and commissioning of MEP development solutions, as well as fire shield, information technology, safeguard and conservation. It was developed as a model to data center design and functionality. Assessment/consistency ratings are revealed in classes 0–4 and are certified by experts who are BICSI-certified [9].

4.6.3 ANSI/TIA 942-A 2014

Business Information Technology: This model excels for IT cabling and network guidance and offers a wide range of issues and troubleshooting opportunities. Trust in the Uptime Foundation process. In the year 2013, UI demanded that TIA halt utilizing the Tier system to specify the level of reliability, and that TIA change to use the word "Rated" instead of "Tier," named 1–4. TIA utilizes a well-integrated set of products to easily identify communications, design, electronics and equipment. Below is a sample of the 2005 model (Table 4.2).

4.6.4 An International Standard: EN 50600

International model light model, for example: The series of EN 50600 is constantly evolving. Many features of this standard reflect TIA, UI and BICSI standards. The criteria are supported by candidates from 1 to 4. Standards are divided into the following sections:

- EN 50600-1 Width ideas
- EN 50600-2-1 Construction structures
- EN 50600-2-2 Energy plants
- EN 50600-2-3 Ecological controls
- EN 50600-2-4 Infrastructure telecommunication cables

TABLE 4.2
Reference Guide

	TIER 1	TIER 2	TIER 3	TIER 4
ELECTRICAL				
General				
Number or delivery Paths	1	1	1 active and 1 Passive	2 active
Utility Entrance	Single feed	Single feed	Dual feed(600 volts or higher)	Dual feed (600 volts or higher from different utility substations)
System allows concurrent maintenance	No	No	Yes	Yes
Computer & Telecommunications Equipment power Cords	Single cord feed with 100% capacity	Dual cord feed with 100% capacity	Dual cord feed with 100% capacity on each cord	Dual cord feed with 100% capacity on each cord
All electrical system equipment labeled with certification from 3rd party test laboratory	Yes	Yes	Yes	Yes
Single points of Failure	One or more single points of failure for distribution system serving electrical equipment of mechanical system	One or more single points of failure for distribution system serving electrical equipment of mechanical system	No single points of failure for distinction system serving electrical equipment or mechanical system	No single points of failure for distinction system serving electrical equipment or mechanical system

Data Center Security

Critical Load System transfer	Automatic Transfer Switch (ATS) with maintenance bypass feature for serving the switch with interruption in power automatic changeover from utility to generator when a power outage occurs	Automatic Transfer Switch (ATS) with maintenance bypass feature for serving the switch with interruption in power automatic changeover from utility to generator when a power outage occurs	Automatic Transfer Switch (ATS) with maintenance bypass feature for serving the switch with interruption in power automatic changeover from utility to generator when a power outage occurs	Automatic Transfer Switch (ATS) with maintenance bypass feature for serving the switch with interruption in power automatic changeover from utility to generator when a power outage occurs
Site switchgear	None	None	Fixed air circuit breakers of fixed molded case breakers Mechanical interlocking of breaks any switchgear in distribution system for maintenance with by passes without dropping the critical load	Drawout air circuit breakers of fixed molded case breakers Mechanical interlocking of breaks any switchgear in distribution system for maintenance with by passes without dropping the critical load
Generators correctly sized according to installed capacity of UPS	Yes	Yes	Yes	Yes
Generator fuel capacity (at full load)	8 hrs (no generator required if UPS has 8 minutes of backup time)	24 hrs	72 hrs	96 hrs

- EN 50600-2-5 Security systems
- EN 50600-2-6 Operations and information management

4.6.5 Governing Standards

Industrial public info institutions are regulated and may be HIPPA (Health Insurance Liability and Supply Act), SOX (Sarbanes Oxley) (2002), SAS 70 Type I or II, GLBA (Gramm-Leach Bliley Act) and new rules may vary subject to the nature of business and current security status in place.

4.6.6 Operating Standards

There exist many performance standards to decide from. Here are the templates that steer your daily routine and procedures as you set up your data center:

- Uptime Institute: Operational Sustainability (with and without Tier certification)
- ISO 9000—Quality System
- ISO 14000—Environmental Management System
- ISO 27001—Information Security
- PCI—Payment Card Industry Security Standard
- SOC, SAS70 & ISAE 3402 or SSAE16, FFIEC (USA)—Assurance Controls
- AMS-IX—Amsterdam Internet Exchange—Data Centre Business Continuity Standard
- EN 50600-2-6 Management and Operational Information

4.7 GREEN DATA CENTER

Power management is becoming an increasingly important issue for Internet services supported by multiple geo-distributed data centers. These data center's energy consumptions and costs are becoming unacceptably high, and placing a heavy burden on both energy resources and the environment [10]. When IT and foreign companies implemented "green environmental protection" systems around the world, we raised issues such as green computers and green storage. He said a small mail produced about 4 grams of CO_2 equivalent. The Green Data Center was born to reduce greenhouse gas emissions and prevent the harmful effects of carbon emissions [11, 12].

4.7.1 Significance of Green Data Centers

Minimal energy efficiency and environmental impact are main focus for green data centers. The warehouses used to store, manage and distribute data (all systems, including energy-saving mechanical and electrical systems) are known as green or sustainable data centers. Reduce carbon monoxide to save a lot of cost and improve efficiency. These storage facilities help modern businesses save energy and reduce fuel consumption. On a global scale, their use is increasing in large- and medium-sized companies. From collection to production, testing to export, these data centers can be used for multiple purposes.

The green data market is estimated to expand at a steady CAGR of around 27% by 2024. This remarkable growth is attributed to the economic growth in the Asia-Pacific region in recent years. The region is likely to grow steadily over the years of events due to the emerging markets in countries such as India and China [13].

4.7.2 Grading Environmental Sustainability

Regardless of industry, environmental protection has become a small industry. Global warming due to carbon dioxide, rising oceans and pollution is making communities work together to find solutions. Accountability is achieved by establishing objectives and exhibiting positive results. In the IT industry, lowering fossil power generation is key, shadowed by water preservation and waste administration. Site data center is one of the major per capita water consumption centers [14]. The potential power of a data center is the size of a small town and requires large amounts of water for cooling [15].

4.7.2.1 Rapid Data Growth Thriving Power Demand

IDC forecasts that the world's data will increase from 33 zettabytes in 2017 to 175 zettabytes in 2025 and the operational costs of these centers will continue to double up every four years, meaning that they need a rapidly expanding arc in the IT sector. Technological progress is unpredictable, but many models predict that by 2030, central energy consumption data could exceed 10% of the world's electricity supply.

Of all these factors, building efficient, sustainable, multi-site data centers is of paramount importance to the environment and business. The green data center is based on a commitment to new green and renewable platforms involving green power, water restoration, cold water, reprocessing and waste administration. They do not include outdated systems and no deposits of new and improved technology.

4.7.2.2 Economic Scalability

Economic Scalability is very important. Instead of companies striving to provide green IT space for service, it creates a green data center that rolls into the haystack at a cost which is less and better. Then, retention benefits are passed on to all or any customers who use its services, and many such businesses can be accessed through a green data center. Additionally, once you interact with a real green data center that uses the activation mode for a long time, the result far outweighs the requirement for green energy. It can be attributed to the environment and benefits of outsourcing IT foundation.

The finest green data center operatives have begun to record and communicate their growth in their annual environmental, sustainability and management reports. For traditional businesses and data centers, sustainability can be measured as part of their management in future.

QTS Realty Trust is among the few data companies responsible for the global community and adheres to the most effective and timely best practices. Finally, the data performance standards for the next few years are determined. QTS encourages the reduction of high carbon content in fuel, water recycling and recycling through the application of continuous and renewable energy technologies, through continuous support.

4.7.2.3 Guidelines for Selecting a Green Data Center

For website workers, information about transitioning to a green data center, or for organizations that already apply for topic green sites, here are nine tips for measuring information center:

- Check ESG "components" and contact GRESB, Carbon Disclosure Project, RE100 and Sustainalytic and other organizations for written data on 100% renewable energy. 40A4 total cost and electricity consumption.
- Check the status of the EPA to obtain the data points that led to the power management.
- Look for no cold water solutions in the industry with 100% fresh air and solar energy.
- Renewable energy must have a significant impact to be effective. Look for data centers that have new energy purchases to get them to buy renewable energy at one price or less than the energy typically produced.
- Uncover new, data-driven service distribution models that use artificial intelligence, machine learning and analytical analysis that can support maintenance planning.
- Find moderate office data that works well for power consumption to create rates and regulations that render it easier and more rewarding for people to buy new energy.
- Search for new suppliers, such as "QTS Growth," which plants more than 20,000 trees in the Sierra Leone Mountains to represent its customers each year, or "Single" People's Services that promotes clean water solutions for new industries.
- Identify clients who speak to and work with prominent organizations such as the EPA Green Energy Partner, REBA, the Information Center Data Committee and the RE100.
- Search for companies that demonstrate networking features such as heat treatment and lighting, recycling, recycling, and waste and EV facilities.

4.7.3 GREEN DATA CENTER BENEFITS

4.7.3.1 Further Testing of Biological Diversity

The BFSI team and other industries have seen knowledge grow over the years. The rapid spread of information has prompted companies to search for shared data centers. Such data centers are more advanced and less expensive than traditional data centers to meet corporate data requirements. Storage areas can be used to move data from any business model and size.

4.7.3.2 Minimal Impact on the Environment

Compared to traditional data centers, green data centers provide facilities to reduce energy consumption and reduce environmental impact. In addition, location information is easily met with new equipment and new energy-saving methods. This type of exercise helps to reduce carbon footprint and benefit the environment.

Data Center Security

4.7.3.3 Consumes Less Power

It improves energy efficiency. Virtualization enables IT staff to monitor and control equipment from remote locations. The main lighting system allows optimum heat dissipation with low light intensity. Even a small temperature rise can reduce energy costs, and a data center allows operators to maintain the temperature while reducing energy.

4.7.3.4 Closing the Unused Servers

Traditional areas will always allocate a particular area according to the needs of the users. After some time, when demand increases, the data center will automatically allocate additional space. In the meantime, this program is always building a place where workers die. In terms of feasibility, even five of all services are not promoted or overused for that purpose. However, these servers use energy and other resources to expand the value. Green data centers can shut down this service and reduce energy consumption and costs.

4.7.3.5 Lower Financial Costs

One of the main reasons one has to pay so much for using a data center is that traditional data centers consume a lot of power. These green or eco-friendly data sources use minimal energy, thanks to constant monitoring and improved data efficiency. In addition, this site data can be retrieved from unused energy for use by various programs. However, the descriptions of this place are better, which is why they are so expensive.

Virtual and dark data centers are very efficient and consume less energy than traditional data centers. Information site management can help your business take full advantage of such environmentally friendly data centers.

4.8 DATA CENTER SECURITY

Organization may currently have strong IT regulations, processes and systems, but could it be prepared for what's coming? Early alert and diagnosis of breaches are decisive to being in circumstances of readiness, and therefore the emphasis of cybersecurity has altered to threaten cleverness [16]. Data center involves complex process; security equipment must be determined separately but at the same time must comply with security regulations. Security is always divided into physical security and software security. Physical protection includes performance measures and procedures needed to prevent external interference. Software or virtual security can bypass firewalls, crack passwords or other intrusions to prevent cybercriminals from entering the network [17].

4.8.1 Physical Protection

The most obvious is the security component of the data center associated with design and organization. The building itself can also be designed as a built-in or multi-purpose unit, which serves as a common space and should house companies that are not connected to the data center. Data center buildings are typically built away from major roads to define a buffer zone of landscape features and collision barriers.

It allows access to a light source. Most do not have external windows or similar access points. Inside guards monitor suspicious activities using surveillance camera images attached to the perimeter of the surface. Guests can use dual authentication to access the building, including a PIV scan and a personal access code. Employee reading and problem statistics (such as fingerprints, iris scans and face details) are not allowed.

4.8.2 Securing the Software

Hacking, malware and spyware are known threats to data stored in archives. The Information Security and Management System (SIEM) provides a clear overview of the safety conditions of the site. SIEM helps provide visibility and control of everything from access, alarm systems and peripheral phone sensors.

Creating a secure network environment is a way to set up a data security. Leaders can divide the network into three areas: the most flexible test site, the rigorous design area and the production area as a development tool.

Before publishing the program and code, the hassle may be the ease of use in some equipment and then providing measurement and repair capabilities. The code can also be run on the lock to check for imperfections or other security measures. With cloud information, the detection of potential data channels may require malware that can be encrypted to control overcrowding.

4.9 CRITICAL ASPECTS FOR SECURING DATA CENTER

Let us appraise 5 best examples for increased fitness and safety. It is unfeasible to list all the available skills and steps, so we will focus on the most vital and important things.

4.9.1 Safety Information

When it comes to the immune system, there are many things to ponder. Another main concern is a home or office building. The structure may have a single purpose, focus on data center acquisition or may have tasks and offices other than the data center. Now we can easily say that the first is the best for maintaining your information securely. The facility's location is usually secluded, with a small number of windows and walls that are not bulletproof to protect it from outside threats, the environment or others. Other key physical safety features comprise 24/7 video inspection, security sentinels and metal technologies and cast-off safety measures, adapted to reflect data sensitivity, safeguarded data, security turnpikes, limit or single source of response.

4.9.2 Monitoring and Authorization Access

Personal error also poses a risk to security, as does storage. A secure environment, especially those responsible for servers and critical assets, should not be

Data Center Security

compromised by unauthorized personnel. To this end, the knowledge center requires manifold access mechanisms across all bodily and digital layers. Access cards and tags are the first measure to consider; even home offices cannot access them. Other precautions include regular inspections by legitimate personnel, pressure as the guest enters and exits the area, and biometric locks. Biometric technology is a high-quality security layer that relies on a person's unique characteristics (such as fingerprints and cartilage). In addition to traditional access control cards, more and more organizations are using biometric courses. Based on knowledge of the materials and equipment mentioned, it should maintain special protection for each room and area. Each security zone requires a high level of validation and regulatory access, as most employees do not have access to all parts of the information center.

4.9.3 Improving Network Security

Ensure perimeter, firewall and data access (IDS) security is in place to help monitor congestion before communication occurs. Libraries often use managed domain names (ACLs) to strengthen their protection. New devices have come with ACLs to allow or block traffic in specific locations by scanning header documents. By building an ACL to establish a road safety engine, you can only allow special technical support. You want to set up the ACL on the side path and hostname. Check the firewall. Unauthorized protection is the primary purpose of any rocket, and its role is due to the first line of security of every website that is safe and secure. In addition, during access control, one should keep in mind to monitor IP addresses and provide a variety of security measures. What about the various unimportant warnings issued by security agencies? It will be a way to look at the crowd from the crowd and make a lot of noise on the ground.

Systems detection of interference is one more critical aspect of network security in the data center. They can identify any unusual activity or signs of fraud, DDoS attacks and other common or major threats.

Two-point and three-point authentication can serve as security protection throughout the network. The information center should set up a redundancy probation team once or twice a year, and it is best to have a certified third party do a pen test. The Zero

Efficiency Standard is important for data center security and applies to everyone entering the building and everyone in the vehicle, so it's easier to see inside both ends even small dangers.

4.9.4 Data Protection

Data security is inseparable from data security. In order to properly preserve and store data, all information must have maximum protections during the broadcast, checked and protected multiple times in less time. In addition, all data security methods are subject to the latest developments, techniques and technologies. Strong legal policies and a healthy sense of cybersecurity of information-connected employees need to be included.

4.9.5 Infrastructure Redundancy

Regarding the data development sector, we have mentioned the infrastructure projects it uses. Because the data center has an important foundation needed to run an organization, downtime can be a very important factor in data center security. All events should take place with tenderness. Cooling is very important for short-term equipment improvement because high technology works the way it heats up. Excessive heating can cause damage to equipment and any data center needs a good security policy. Crimes can occur for a variety of motives, from actual ride out to human error. It may also be due to a power failure or electric shock. However, the UPS must be in the storage area and other emergency equipment.

4.10 DATA CENTER SECURITY STANDARDS

Here are some of the key body safety data security technologies and models that tenants should consider if they want to interact with a website.

4.10.1 Safety Level

Every aspect of the safety of light centers should be linked to other aspects as part of a complete waste management system. The point is that the intruder will be forced to break some kind of security before accessing sensitive data or valuables in the locker room. If one song fails, the other one can prevent intrusion from destroying the entire system.

4.10.2 Logging in Access

While this may seem like a simple matter, one of the most important aspects of security in the knowledge area is to ensure that only authorized persons have access to equipment. When an organization has an awareness center, not all employees have access to services. This remains an integral part of Zero Trust's security philosophy. By monitoring access logins, clinics can help their clients protect themselves from theft and prevent human error from unauthorized IT product operators from the start.

4.10.3 Video Surveillance

Another enduring physical safeguard of digital technology is video surveillance, which is also useful for data storage. For the sake of information quality, you need a surveillance camera (CCTV) with full sound image, face tilt and zoom function to monitor everyone outside and inside the door itself. Photos should be digitally protected and stored on the website to prevent unauthorized access.

4.10.4 Security Access Protocol

Functional areas such as the level of information should be protected with very simple locks. Custom lists or traps can prevent licensed visitors from issuing certificates to others, which is the standard for protecting any data repository subject.

Data Center Security

4.10.5 Security Round the Day/Year

Security checkpoints, cameras and alarms will not be available if there are no personnel on site to respond to threats and unauthorized actions. Regular police officers in each area of the information centers are often reminded that security guards are alert and can respond quickly to deal with any problems that may arise.

4.10.6 RFID Asset Management

Although it is important to have secure central personnel in the data center and camera storage, it is still difficult to monitor at least all computer equipment. With RFID tags, data centers can use powerful business software to manage and monitor assets in real time. It can also send instant alerts upon downloads or deliver assets, allowing information personnel to respond quickly to threats.

4.10.7 Employee Verification

Check whether the type of employees varies between employees and remote technicians in data center security facilities. Internal auditors and implementing the appraisal by all external auditors can trust their customers that they can trust and manage their assets.

4.10.8 Relieving Plans

If someone has access to areas and sensitive features of the information center but maintains his or her position, he or she will not be able to access them. Whether it be office data or user resources and entities accessed by the organization, facilities must have systems in place to eliminate those time constraints. This may include updating access lists, collecting keys or removing biometrics from the facility system to ensure that they are not ready in the future.

4.10.9 Multi-Factor Authentication

All data centers are subject to "fake" security procedures which must include a variety of tests. Each entry requires two or more types of notices or permits to ensure that no one is "softened" by security if they lose some form of admission.

4.10.10 Biometrics

Biological identification technology is an advanced technology in the field of security that allows people to identify physical features such as thumb, retina or sound structures. There are many ways to integrate biometric technology into legal spaces, and its specific role plays a role in both proofs.

With the continuous participation of data center security technology, it is inevitable that new physical security measures will be adopted as best practices. First, data center data protection models may not be visible, because most of them are doomed to fail. However, consumers of data centers can review security certificates and request a more detailed summary of the physical and psychological security measures put in place to ensure that data is properly protected.

4.11 CASE STUDIES

4.11.1 Microsoft Green under Water Data Centers

The Microsoft project team Natick used the Nordic Islands data center 117 feet above sea level in the spring of 2018. Over the next two years, team members will test and monitor data center server performance and reliability [18] (Figure 4.4).

4.11.1.1 Proof of Concept

An underwater data center concept was raised at Microsoft in 2014 during a Thinkweek event, a time to gather staff for dissemination of external ideas. The idea is realized by providing coastal residents with fast-moving services in the cloud and saving energy.

More than half of the world's population lives within 120 miles of the coast. By placing underwater data centers near coastal ports, the data maintains short distances, resulting in fast, smooth surfing and stream of images. Cool surfaces provide the energy efficient design. For example, they will charge heat exchangers like those found in submarines. Natick from the Microsoft team confirmed that the underwater data vision was possible during the 105-day Pacific operation in 2015. Phase II of the project involved contracts and sea freight, ship building and technical renewable energy experts to demonstrate this idea as an optional extension.

FIGURE 4.4 Microsoft underwater data center.

Data Center Security   75

4.11.1.2 Power Wash and Data Collection

After detecting by sea and before sending it to Orkney Islands, the Green Marine Group cleaned up the pipes located on the servers of the North Islands 864 servers and other air-conditioning industries (Figure 4.5).

The researchers then placed the test tube in a valve at the height of the container to collect an air sample for analysis at Microsoft's headquarters in Redmond, Washington. He also said that the question of how different gases come from wires and other equipment can change the environment of the computer. The archive and wind turbines are mounted in trucks and are operated by Nigg Energy Park, the International Energy Agency in northern Scotland. There, the shipping team released the result and slipped the staff shelf, while Fowers and his team made health checks and took the items to be sent to Redmond for testing. Of the items collected and shipped to Redmond, there were incomplete pairs for the wiring and wiring harnesses. Researchers believe that these devices help them understand why workers spend eight hours more time working on water than they do on the ground.

The air group is thought to contain nitrogen, it is much smaller than iron and, therefore, the absence of contaminants is the main reason for some. If the analysis proves to be thorough, the team can also prepare to convert the findings into a ground-based data center. Members of the project from Natick have cleaned up the North Island Underwater Records Center, collected from the sea off the Orkney Islands in Scotland. In the two years under the sea, there is an opportunity to form thin and shiny mosses in the pipes, and ocean anemone grows in large cantaloupe in the shade of deciduous trees.

FIGURE 4.5 Power washing data center.

4.11.1.3 Energy, Waste and Water

Other courses in the Natick project describe discussions on improving data centers for scientists and researchers. Cutler considers issues such as integrating underwater data into wind turbines. Even if the wind is light, the data center medium is strong enough. According to the final test, the required fiber-optic cable for mobile data will be used to calculate the power connected to the coast. Other care benefits may include removing the need for replacement. In the broadcast settings, all servers are replaced every five years. High server reliability means getting an early server.

Additionally, Natick project has shown that location data that is normally run and kept safe without cutting off water is critical to people, farms and animals, Cutler says.

REFERENCES

[1] Louise Krug, Mark Shackleton and Fabrice Saffre, "Understanding the Environmental Costs of Fixed Line Networking", Proceedings of the 5th International Conference on Future Energy Systems, pp. 87–95, 2014.
[2] Bernadette Johnson "How Data Centers Work" 27 October 2013. HowStuffWorks.com. https://computer.howstuffworks.com/data-centers.htm
[3] Alex Sharp "Building the Data Center of the Future" 23 July 2020. Datacenterfrontier. com https://datacenterfrontier.com/building-data-center-of-the-future
[4] Alan Seal "What You Need to Know About Data Center Design Standards" 21 May 2019. vxchnge.com www.vxchnge.com/blog/data-center-design-standards
[5] ServerLIFT Corporation "Modern Data Center Design and Architecture" 15 January 2020. serverlift.com https://serverlift.com/blog/modern-data-center-design-and-architecture/
[6] Data Center Infrastructure "Physical Components of a Data Center" digitalrealty.com www.digitalrealty.com/data-center-infrastructure
[7] S. Qi, Y. Zhang and M. Wang, "Study and Application on Data Center Infrastructure Management System Based on Artificial Intelligence (AI) and Big Data Technology", 2019 IEEE 4th International Future Energy Electronics Conference (IFEEC), 2019, pp. 1–4.
[8] Smart Sheets "Reduce Costs and Improve Performance with Data Center Infrastructure Management" smartsheet.com www.smartsheet.com/all-about-data-center-infrastructure-management>
[9] Steven Shapiro "Data Center Design" 6 January 2016. datacenterknowledge.com www.datacenterknowledge.com/archives/2016/01/06/data-center-design-which-standards-to-follow
[10] Rahman, Xue Liu and Fanxin Kong, "A Survey on Geographic Load Balancing Based Data Center Power Management in the Smart Grid Environment", *IEEE Communications Surveys & Tutorials*, 16, 1, 214–233, 2014.
[11] Silver Touch "How Green Data Center Benefits Modern Business and Environment" silvertouch.com www.silvertouch.com/how-green-data-center-benefits-modern-business-and-environment/
[12] A. Roy, A. Datta, J. Siddiquee, B. Poddar, B. Biswas, S. Saha, P. Sarkar, "Energy-efficient Data Centers and Smart Temperature Control System with IoT Sensing", 2016 IEEE 7th Annual Information Technology, Electronics and Mobile Communication Conference (IEMCON), 2016, pp. 1–4.

[13] C. Peoples, G. Parr, S. McClean, B. Scotney, P. Morrow, "Performance Evaluation of Green Data Centre Management Supporting Sustainable Growth of the Internet of Things", *Simulation Modelling Practice and Theory*, 34 (2013) 221–242.
[14] D. Gmach, Y. Chen, A. Shah, J. Rolia, C. Bash, T. Christian, et al., "Profiling Sustainability of Data Centers", 2010 IEEE International Symposium on Sustainable Systems and Technology (ISSST), pp. 1–6, 2010.
[15] QTS Intelligent Infrastructure "Green Data Centers—Scaling Environmental Sustainability for Business and Consumers Collectively" 28 July 2020.
[16] Dulhare, Uma N. and Shaik Rasool. "IoT Evolution and Security Challenges in Cyber Space: IoT Security." Countering Cyber Attacks and Preserving the Integrity and Availability of Critical Systems, edited by S. Geetha and Asnath Victy Phamila, IGI Global, Hershey, USA, 2019, pp. 99–127.
[17] Sara Jelen "What is Data Center Security? Top 5 Best Practices" SECURITYTRAILS BLOG, 3 December 2019.
[18] John Roach "Microsoft Finds Underwater Datacenters Are Reliable, Practical and Use Energy Sustainably", Microsoft Innovation Stories, 14 September 2020.

5 Energy-Efficient Network Intrusion Detection Systems in the IOT Networks

Alka Singhal, Bhushan K Jindal and Veepsa Bhatia

CONTENTS

- 5.1 Introduction 80
- 5.2 IOT Architectures 80
  - 5.2.1 Three-Layer IOT Architecture 81
  - 5.2.2 Five-Layer Architecture 81
  - 5.2.3 Six-Layer Architecture 83
  - 5.2.4 Fog-Based Architecture Layers 83
- 5.3 IOT Applications 83
  - 5.3.1 Wearable Healthcare 83
  - 5.3.2 Smart Home Applications 84
  - 5.3.3 Smart Cities 85
  - 5.3.4 IOT-Based Agriculture 85
  - 5.3.5 IOT-Based Industrialization 85
- 5.4 Security Threats in IOT Applications 86
  - 5.4.1 Literature Survey 86
  - 5.4.2 Security Issues in IOT 86
    - 5.4.2.1 Security Challenges for IOT Devices 86
    - 5.4.2.2 Security Threats in IOT 86
  - 5.4.3 Existing Security Mechanisms in IOT 90
    - 5.4.3.1 Intrusion Detection Systems 90
    - 5.4.3.2 Methods of Intrusion Detection 92
    - 5.4.3.3 Security in Internet of Things Using Intrusion Detection Systems—Proposed System 92
    - 5.4.3.4 Energy-Efficient Security 93
  - 5.4.4 Conclusion 93
- References 93

DOI: 10.1201/9781003097198-5

5.1 INTRODUCTION

IOT can be defined as a network, where things, including embedded devices, are interconnected. In IOT, an interconnected network is used with remotely controlled sensors to perform functionality. The interconnected sensor devices vary in sizes to provide desired tracking and analysis to get correct results. Some of the examples are wearable watches monitoring patients pulse rate, BP and remotely controlled home appliances. According to McKinsey: "Sensors and actuators embedded in physical objects are linked through wired and wireless networks, often using the same Internet Protocol (IP) that connects the Internet." As the number of smart devices are increasing, the IOT is creating a wide demand in the market with available lossy networks and limited resources.

> The Internet of things will involve a massive build-out of connected devices and sensors woven into the fabric of our lives and businesses. Devices deeply embedded in public and private places will recognize us and adapt to our requirements for comfort, safety, streamlined commerce, entertainment, education, resource conservation, operational efficiency and personal well-being.

"Rise of the Embedded Internet." The IOT has wide scope from medical assistance to weather forecasting, traffic surveillance, etc. It has many applications like smart automated homes with remotely automated lights, appliances, door, locks, smoke detector, wearable smart watches, smart cities with automated traffic lights, water distribution, waste management, self-driving cars, smart inventory management and smart farming.

Singh et al. [1] state that IOT will be covering many day-to-day requirements soon. As IOT is a collection of heterogeneous devices using the conventional available networks based on unreliable-natured IP protocol, the IOT also faces the same security issue as other network applications. An additional challenge faced in developing an IOT application is the availability of constrained resources that are limited power and memory. The deployed IOT network is very much prone to the attacks which are a threat to privacy and confidential data. It is found in my studies that IOT is vulnerable to attacks. The major reason is its complexity; to make an IOT device operable, there are various layers involved; those are devices, communication channels, cloud interfaces and application surfaces. Each of these layers is susceptible to attacks making it more prone to attacks. Just like communication channel can be attacked with an open network traffic, cloud is intervened by inadequate passwords and false credentials. Real-time Analytics with constrained memory and power issues exposes the system for network attacks. Therefore, there is a big requirement of security solutions for the constrained environment. In literature survey, we are discussing a brief of the available security solutions for IOT environments. The literature survey will cover all types of security breaching possible in IOT architectures.

5.2 IOT ARCHITECTURES

A simple IOT device is composed of embedded sensors, actuators, networks, processors and transceivers. Sensors and actuators interact with the physical environment. Sensors collect the data which is stored and provide useful inferences.

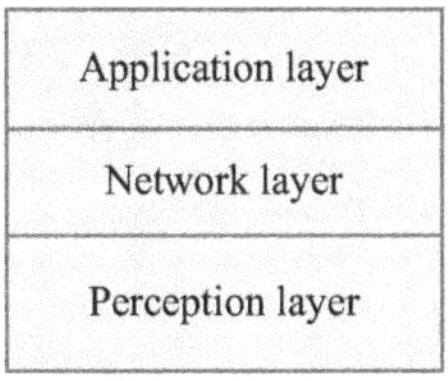

Network Intrusion Detection Systems 81

| Application layer |
| Network layer |
| Perception layer |

FIGURE 5.1 Three-layer architecture.

Actuators, however, effect a change in environment in response to inferences, e.g., a temperature controller. The storage and processing at the other end of the network known as server. The pre-processing can be done at the either edge: transmission side or receiver side. As explained earlier, the IOT devices work in very resource-constrained environments like small size, low energy, power and computational capabilities. At the initial stage when IOT devices were evolving, there was big research in getting desired accurate results with the available resources. Greatest challenges were data collection, storage and communication. For communication, IOT devices used available wireless networks. Sensors, actuators, compute servers and the communication network form the core infrastructure of an IOT framework.

Burhan et al. [2] explain various architectures for IOT involving a variable number of layers in order to get the best, fast, reliable and secure convergence of information. Number of layers depends on requirements and tasks to handle. The most basic architecture available for IOT is the three-layer architecture (Figure 5.1).

5.2.1 THREE-LAYER IOT ARCHITECTURE

The layers for the three-layer architecture are as follows:

(a) Perception layer: It is the layer which interacts with the physical environment; it is composed of sensors which gather data for further processing. It is the most important layer which is responsible for sensing physical parameters on which processing can be done.
(b) Network layer: It is responsible for transmitting the collected data safely and correctly to the receiver server.
(c) Application layer: It is responsible for application-specific services to the users.

5.2.2 FIVE-LAYER ARCHITECTURE

The five-layer architecture consists of the following:

(a) Perception layer
(b) Transport layer
(c) Processing layer
(d) Application layer
(e) Business layer

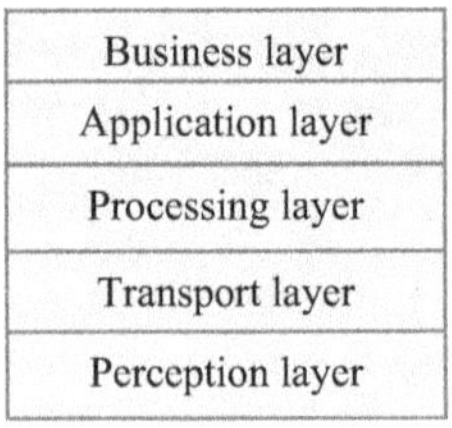

Business layer
Application layer
Processing layer
Transport layer
Perception layer

FIGURE 5.2 Five layers for the three-layer architecture.

Application layer
Sensing layer
Communication layer
Service layer
Infrastructure layer

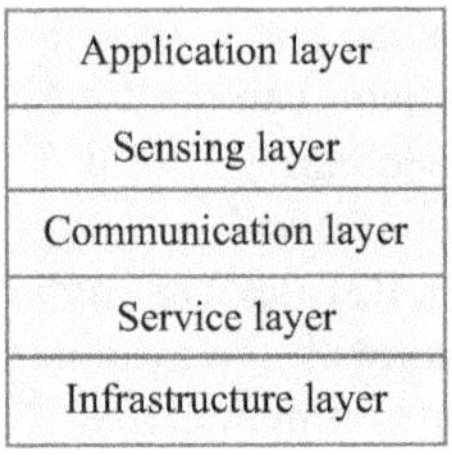

FIGURE 5.3 Modified five layers for the three-layer architecture.

Focus layer
Cognizance layer
Communication layer
Application layer
Infrastructure layer
Competence business layer

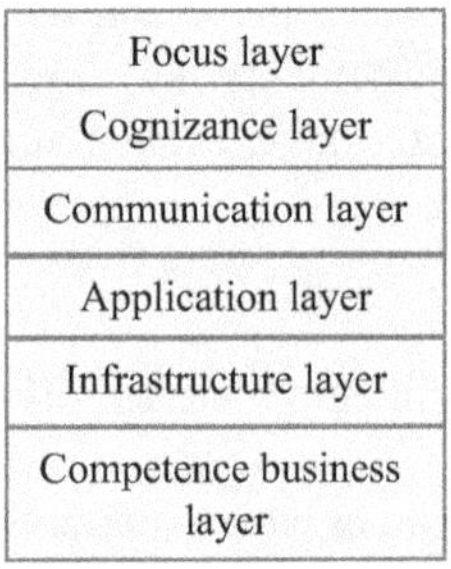

FIGURE 5.4 Modified advanced six layers for the three-layer architecture.

The perception and application layers have the same role as in the three-layer architecture (Figure 5.2). The transport layer is responsible for data transportation from perception layer to processing layer via available networks. The processing layer is the layer which stores, analyzes and processes the data, and provides important inference to the lower layers. This is illustrated in Figure 5.3.

In a modified IOT architecture, the application layer collects the information about the services to be provided to the client. The sensing layer is composed of the sensors and electronic devices. The communication layer helps in data transmission. The service layer helps to perform activities required by clients and finally the infrastructural layer creates GIS mapping, cloud computing and computing storage facilities (Figure 5.4).

Network Intrusion Detection Systems

5.2.3 Six-Layer Architecture

In a six-layer architecture, the focus layer helps to identify the nodes with various aspects; the cognizance layer demonstrates the sensing capability of the information from the objects focused by the focus layer. It is composed of sensors, actuators and data monitoring systems. The transmission layer helps in transmitting data, the application layer helps in collecting and categorizing the information as per requirement, and finally the infrastructure layer provides cognitive computing storage facility and an additional added layer. Competence business assesses the business and profit model with privacy.

5.2.4 Fog-Based Architecture Layers

A Fog-based architecture contains a monitoring layer which monitors power issues, resources and a pre-processing layer performing processing and sensor data analysis. The storage layer provides data replication, distribution and storage (Figure 5.5).

5.3 IOT APPLICATIONS

The IOT applications are in various domains like:

- Wearable and Healthcare
- Smart Home Applications
- Smart Cities
- Agriculture
- Industrial Automation

5.3.1 Wearable Healthcare

There is a huge demand of healthcare devices which are connected to patients and can provide live monitoring of the patient with low operational cost (Figure 5.6). Therefore, it generates a huge demand of IOT in fulfilling this demand. In the period of corona pandemic, lots of research was made on the development of IOT devices which can remotely monitor various health parameters of the affected person. Research was also

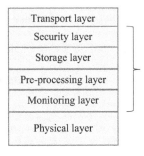

FIGURE 5.5 Fog-based architecture layers.

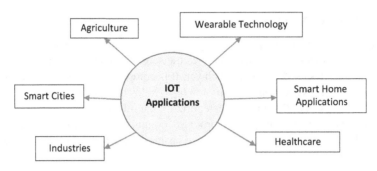

FIGURE 5.6 IOT applications.

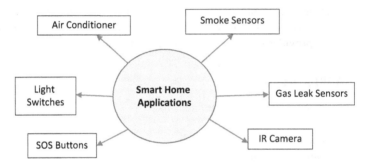

FIGURE 5.7 Smart home applications.

made on the device which can monitor his movements and control it as the disease is highly communicable. A wide range of wearable healthcare devices are coming in the market to assess fitness factors to track a person's day-to-day health goals. Therefore, IOT applications have wide acceptance in Wearable and Healthcare.

5.3.2 SMART HOME APPLICATIONS

In smart homes, IOT plays the role of a key component to make it smart and intelligent (Figure 5.7). These devices are connected within network and share data over home IP network. The home gateway controls the incoming and outgoing flows of information. These smart homes can be viewed as two components:

(a) The sensor devices (which contain sensory hardware with an RF transmitter receiver).
(b) The interface device (which does processing, data collection and communication over internet).

Some of the smart applications are IOT smart devices, which include (1) Cameras, (2) Infrared Sensors, (3) Magnetic Door Sensors, (4) Smoke Sensors, (5) Gas Sensors and (6) Automatic Doors and Windows.

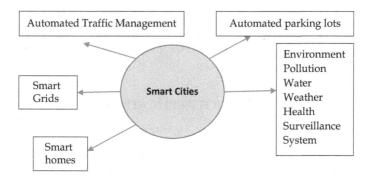

FIGURE 5.8 IOT-based smart cities.

5.3.3 SMART CITIES

Smart cities are those that make the use of these smart things to carry out various functions such as lighting, traffic control, connecting multiple cities, energy consumption and pollution control. The main purpose of smart cities is to replace the way how we look at things. Regarding many aspects where IOT is set to rule, we can say that from the most reliable day-to-day actions to the most complex human emotions, IOT will affect it all. Commonly, from the smart city applications and the underlying environment, the citizens will be benefitted (Figure 5.8).

5.3.4 IOT-BASED AGRICULTURE

IOT in agriculture uses special equipment, wireless network and IT services to provide smart farming. It enables farmers to monitor their fields from anywhere. It enables them to take viable solutions if soil moisture goes down. IOT can also provide automated irrigation in emergency. IOT smart farming is a system built for monitoring the farms using sensors and automated irrigation system.

IOT has huge potential in the form of agricultural drones, smart greenhouses, predictive analysis of smart farming, geofencing, etc.

5.3.5 IOT-BASED INDUSTRIALIZATION

IOT technology can be utilized in industries in the following ways:

(a) Digitized factory containing IOT-enabled machinery that can transmit useful information to the manufacturers and field engineers. It includes process automation and optimization. It will provide key information to the managers on well-defined commands.
(b) Condition-based alerts and maintenance alerts. The IOT-based devices regularly monitor the given parameters like pressure, temperature, and raise alarm if unusual conditions are met. They ensure the smooth working of the equipment in prescribed environment.

(c) The IOT devices provide monitoring of the production lines from start to the packaging. It can provide appropriate steps which can be taken to reduce operational cost. It monitors the inventory and tracks the items on line-item level and gives useful notification.

5.4 SECURITY THREATS IN IOT APPLICATIONS

5.4.1 Literature Survey

Lots of research is being done over security issues in IOT. As IOT applications are highly dependent on network, there is a huge risk of attacks. Razzaq et al. [3] explained that wireless networks are highly susceptible to security breaches. They explored various security protocols and deployment architectures to ensure privacy and security. Rehman et al. [4] included all the major challenges like scalability, surveillance integration in IOT applications, which makes it prone to security attacks. Jing et al. [5] discussed about IOT architectures and its layers and explained about security threats at each layer. They analyzed all the three layers, perception, transportation and application, and found all types of attacks. Hassan et al. [6] did a complete study from 2016 to 2018 to overview the state of the art of IOT security research and relevant tools, IOT modelers and simulators, etc., used to handle it. Fazion and Mauro [7] describe various characteristics like addressability, identification and localization of IOT and find solutions for problems like Identity spoofing, Data tampering and Eavesdropping.

5.4.2 Security Issues in IOT

5.4.2.1 Security Challenges for IOT Devices

It is well understood from the architecture, requirement and applications that IOT devices are resource-constrained. IOT devices are generally low power operated and work on processors of low clock rate. The standard algorithms requiring fast computation and processor requirements cannot be applied in the case of IOT devices. With power and processing constraints, IOT devices also face memory constraints. They have limited RAM and Flash memory. The security solutions must be memory-efficient. They are majorly deployed at remote areas and are easily available for attacks as majorly left unattended. Attacker tampers it by device capture. They easily find out the cryptographic secrets and modify them with malicious codes. The software which is designed for IOT applications has thin protocol stack which is generally unable to handle security breaches. IOT devices cannot be reprogrammed for the new security patch easily. The task of sending the updated code and reinstallation is tough in remotely installed IOT devices. The biggest factor which makes IOT attack-prone is mobility; this raises the need to develop mobility-resilient security algorithms. It is difficult to design a solution for both wired and wireless mediums.

5.4.2.2 Security Threats in IOT

Hossain et al. [8] explain that various IOT security threats such as distributed denial of service (DDoS), ransomware and social engineering can be used to purloin critical

Network Intrusion Detection Systems

data from people as well as organizations. Attackers usually exploit security vulnerabilities in IOT devices and its associated infrastructure for execution of cyber-attacks. These IOT security threats can be more crucial for consumers as they are unaware of existence of such threats and do not have the competence to mitigate them. Hence, organizational leaders need to identify and address such security threats to offer high-end, secure products and services to their consumers.

IOT deployments need to be aware of the following security threats:

(a) Botnets
(b) Denial of Service
(c) Man-in-the-middle
(d) Identity and data theft
(e) Social engineering
(f) Advanced persistent threats
(g) Ransomware
(h) Remote recording

Each of above security threats are explained in brief below (Figure 5.9):

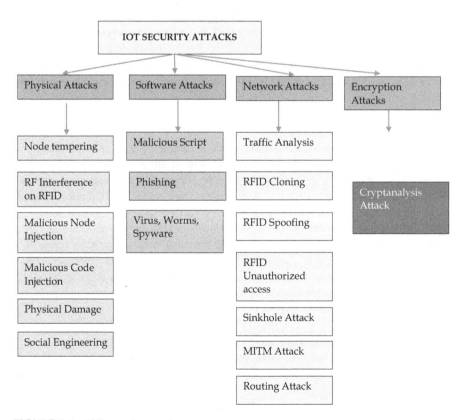

FIGURE 5.9 IOT security attacks.

5.4.2.2.1 Botnets

A botnet is a collection of internet computers called "bots," that are under remote control of some outside party. These botnets remotely take control over a victim's system and distribute malware. Cybercriminals deploy and control botnets using Command-and-Control-Servers to purloin confidential data, capture online-banking data and execute cyber-attacks like DDoS and phishing. Cybercriminals can also utilize the botnets to attack IOT devices that are connected to several other devices such as laptops, desktops and smartphones.

Mirai botnet has displayed how dangerous IOT security threats can be. The Mirai botnet had infected an estimated 2.5 million devices, including routers, printers and smart cameras. Attackers used the botnet to launch DDoS attacks on numerous IOT devices. After witnessing the impact of Mirai, several cybercriminals have developed multiple advanced IOT botnets. These botnets can launch sophisticated cyber-attacks against vulnerable IOT devices.

5.4.2.2.2 Denial of Service

A denial-of-service (DoS) attack deliberately tries to cause a capacity overload in the target system by sending multiple requests. Unlike phishing and brute-force attacks, attackers who implement DoS don't aim to steal critical data. However, DoS can be used to slow down or disable a service to impact the reputation of a business. For instance, an airline that is attacked using DoS will be unable to process requests for booking a new ticket, checking flight status and cancelling a ticket. In such instances, customers may switch to other airlines for air travel. Similarly, IOT security threats such as DoS attacks can badly impact the reputation of businesses and affect their revenue.

5.4.2.2.3 Man-in-the-Middle

Through a Man-in-the-Middle (MiTM) attack, a hacker intrudes into the communication channel between two interacting systems in an attempt to intercept messages among them. Attackers gain control of their communication and send spoofed messages to these systems. Such attacks can be used to hack IOT devices such as smart refrigerators and autonomous vehicles.

MiTM attacks can hence be used to attack various IOT devices as they share their data in real-time. With MiTM, attackers can intercept communications across IOT devices, leading to critical malfunction. For instance, smart home accessories such as bulbs can be controlled by an attacker using MiTM to change its color or turn it on and off. Such attacks can lead to serious consequences for IOT devices such as industrial equipment and medical devices.

5.4.2.2.4 Identity and Data Theft

Several breaches of data were brought to notice in 2018 for compromising the data of large population. Confidential information, of people, such as personal details, bank account details, credit and debit card credentials and email addresses was reported to be stolen in these data breaches. Hackers can now attack IOT devices such as smart watches, smart meters and smart home devices to gain additional data about various users and organizations. By collecting such kind of data, attackers execute more sophisticated and it leads to detailed identity theft.

Attackers also exploit vulnerabilities in IOT devices that are connected to other network devices and larger systems. For example, hackers can attack a vulnerable IOT sensor in an infrastructure and gain access to their business network. In this manner, attackers can intrude into multiple enterprise systems and obtain sensitive business data. So, IOT security threats can lead to a rise in data breaches in multiple businesses.

5.4.2.2.5 Social Engineering

Hackers use social engineering to manipulate people into disclosing their sensitive information such as passwords and bank details. Alternatively, cybercriminals may use social engineering to access a system for installing their malicious software secretly. Usually, social engineering attacks are executed using phishing emails, where an attacker must develop convincing emails to manipulate people. However, social engineering attacks can be simpler to execute in the case of IOT devices.

IOT devices like wearables continuously collect large volumes of personally identifiable information (PII) to provide a personalized experience for users. Such IOT devices use personal information of users to extend user-friendly services, for example, ordering online products with the support of voice commands. However, so-collected PII can also be accessed by attackers to gain confidential information such as details of bank accounts, purchase history as well as user addresses. This information can then enable a cybercriminal to perform an advanced social engineering attack that targets a user and one's family and friends using the vulnerable IOT devices and networks. This way, IOT security threats such as social engineering can be used to gain illegal access to user data.

5.4.2.2.6 Advanced Persistent Threats

Advanced persistent threats (APTs) are a major concern in IOT security for various organizations. An APT is a targeted cyber-attack, where an outsider once gains the illegal access to a computer network and remains undetected for a long duration. Attackers generally aim to monitor network activities and intend for theft of crucial data using APTs. Such cyber-attacks are difficult to prevent, detect or mitigate.

With the emergence of IOT, large volumes of critical data are transferred among several IOT infrastructure devices. A cybercriminal can target such IOT devices to get access to personal or organizational networks. With this approach, cybercriminals can steal lot of important confidential information.

5.4.2.2.7 Ransomware

Ransomware attacks are one of the most notorious information systems' threats. In such an attack, a hacker uses malware to encrypt data, on user device, that may be of great importance for business operations. An attacker will offer to decrypt critical data only after receiving a ransom.

Ransomware can be one of the most sophisticated IOT security threats. Researchers have demonstrated the growing impact of ransomware using smart thermostats. With this approach, researchers have shown that hackers can turn up the temperature and refuse to go back to the normal temperature until they receive a ransom. Similarly, ransomware can also be used to attack IOT devices and smart homes. Similarly, a hacker can attack a smart home and ask the owner to pay a ransom.

After looking at above-mentioned IOT security threats, it can be concluded that there is a great need for inducting a security mechanism in the IOT network infrastructures. Deploying a good security mechanism with well-thought policies can lead to a substantially secure system. Intrusion detection system (IDS) is a widely used security mechanism. The following section gives an insight into IDSs and various types of IDS.

5.4.3 Existing Security Mechanisms in IOT

Figure 5.10 depicts various IOT security mechanisms and the subsections explain in brief about each of these security mechanisms.

5.4.3.1 Intrusion Detection Systems

Anthi et al. [9] explain that an IDS is deployed in a networked infrastructure and monitors the traffic over network with the aim to detect a suspicious activity. The IDSs generate alert whenever any suspicious activity is detected. This is a software

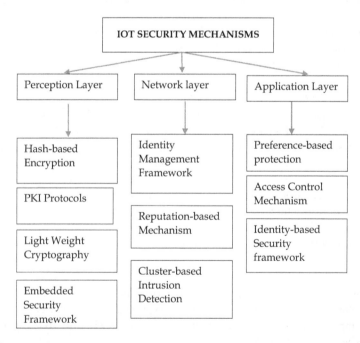

FIGURE 5.10 IOT security mechanisms.

component which is deployed in a network and it scans the network traffic or system for potential breach of policy or any harmful activity. Whenever any such incident is noticed by IDS, the same is either informed to the system administrator or logged through some mechanism like centralized logger. The purpose of the centralized logger is then to collect such incidents/activities from complete network devices and prepare a collective alert mechanism. Apart from the alert mechanism, a formatted report is also generated by the centralized logger of incidents/activities detected for a duration of time.

Benkhelifa et al. [10] explain that the purpose of IDSs is to detect potential harmful activities by monitoring the network. Sometimes, false alarms are also raised by IDSs due to improper configuration. Hence, the network/system administrators need to configure the IDS deployment properly for rules and policies. This configuration involves setting up IDS to recognize what normal traffic looks like from a malicious one.

There are 5 types of IDSs as given below:

5.4.3.1.1 Host Intrusion Detection System

An HIDS monitors the inbound and outbound flows on a network device and sends an alert to the administrator if any malicious activity is detected. This way, HIDS is deployed on individual host or device. This is generally deployed on machines or devices whose file footprint is not expected to change. So HIDS takes a snapshot of the files currently residing in the system and compares the same with the snapshot taken previously. These two snapshots are then compared for detection of any anomaly. If found so, the same is issued as an alert to the administrator.

5.4.3.1.2 Network Intrusion Detection System

Network Intrusion Detection System is put on a boundary of a subnet to analyze the inbound and outbound traffic flows for the subnet. It examines all the data from the connected devices of the network. The data that flows in the network is compared with the one which is expected to flow, and any mismatch is issued as an alert to the administrator so that suitable countermeasures can be taken.

5.4.3.1.3 Protocol-Based Intrusion Detection System

A Protocol-Based Intrusion Detection System is deployed on the servers where most communication is done using specific protocols. This kind of IDS is generally used for HTTP or HTTPS traffic and sits in front of web server providing front end to the user. PIDS inspects the HTTP(s) traffic on a web server with respect to the expected flow as per system implementation.

5.4.3.1.4 Application Protocol-Based Intrusion Detection System

Application Protocol-Based Intrusion Detection System works for a group of servers. In any system, there are protocols at the application level that are well defined. An APIDS will monitor the traffic for this set of servers to identify intrusion based on comparison of defined protocol and data traffic in flow.

5.4.3.1.5 Hybrid Intrusion Detection System

Hybrid Intrusion Detection System is a combination of above given two or more approaches. This kind of IDS is more effective than an individual type of IDS because a single IDS looks at multiple angles of intrusion.

Above are the types of IDS, and below given are methods used in IDSs:

5.4.3.2 Methods of Intrusion Detection

5.4.3.2.1 Signature-Based Intrusion Detection

Signatures of malicious software are the way to identify that what kind of malicious software is trying for intrusion in the system. These signatures are generally a bit patterned with a fixed occurrence of particular order of bit values. Hence, signatures of malicious software that become known are captured in IDS for the detection of existence of such malicious traffic/software.

5.4.3.2.2 Anomaly-Based Detection

All the malwares are not always known. Any new malware will have its own signature which is not yet known. Anomaly-based IDS scheme uses machine learning to learn trustful activity behavior of the system and shall point out any anomalous-looking activity in the system which is not in line with the learnt trustful activity pattern.

Machine learning-based method gives better results as these can be trained according to actual software and hardware behavioral pattern.

5.4.3.3 Security in Internet of Things Using Intrusion Detection Systems—Proposed System

With the advancements in IOT systems, the threats are also increasing manifold for these IOT systems as discussed in the section about threats to IOT systems above. Various security goals of IOT systems are given below:

- **Authentication**: It is all about verifying the true identity of some user or device in the IOT systems. This ensures that a legitimate user or device is getting access to the IOT system.
- **Authorization**: It defines the level of access that needs to be provided based on the authentication credentials provided by the user. Every system works on role/identity-based authorization levels. An authenticated person may be less authorized than another authenticated person who is more authorized. The levels of authorization enable well-defined privileges to access system resources and application features.
- **Availability**: It is explained as the capability of the IOT system to ensure that the system is available, i.e., reachable, accessible and usable whenever needed.
- **Privacy**: Data privacy is the biggest concern in modern-day information systems and Internet of Things-based systems. The goal here is to ensure that only necessary information about the user is kept in the system and it is ensured that it is saved in such a way that it is accessed only by authorized personnel and is protected from all unauthorized access.

- **Integrity**: It is the trustworthiness and dependability of the information in the IOT systems. Any user accessing information about its connected car or smart home should get only correct information. This is also desirable where historical information is stored by the IOT system. It describes that information stored in the system should be dependable and trustworthy even in the future.
- **Confidentiality**: It is the protection of information available in the IOT system. It's a promise of the authorized users of the system that the information marked as confidential in the system is not shared with anybody who is not supposed to access the same.

Looking at the above goals of IOT systems, inclusion of security mechanisms, while keeping these goals the aim of security solution, becomes an urgent need. The user data of such systems shall also be secured with the presence of implemented security. Deployment of IDSs in IOT systems enables manifold security enhancement through techniques discussed in the section above.

5.4.3.4 Energy-Efficient Security

Providing security through IDS for general computer systems, networks and IT infrastructure of a wired network does not face the challenge of managing the power consumed by the security solution or devices. The IOT devices are deployed where communication is over wireless networks and may also be using 6LoWPAN technology. Hence, providing security to such IOT devices needs to be very much energy-efficient. The sensors deployed in remote wireless terrains are generally equipped with low power availability. Foreseeing the need for solutions that consume the least power, it has become the latest research interest to identify mechanisms that can be adopted by IDSs.

5.4.4 CONCLUSION

IOT is a rapidly growing technology and will be incorporated in most of the tasks of daily life. The application varies from general to extremely data-sensitive. Therefore, there is a much requirement of security in IOT systems. In this chapter, we have discussed various challenges and constraints present in IOT systems, which make it attack-prone and less secure. We tried to study all the existing methods available as security solutions for IOT systems and found IDS as one of the holistic methods which tries to cover all types of architectural, network attacks in IOT systems.

REFERENCES

[1] Singh, Sachchidanand, and Nirmala Singh. "Internet of Things (IoT): security challenges, business opportunities & reference architecture for E-commerce." *2015 International Conference on Green Computing and Internet of Things (ICGCIoT)*, 2015.

[2] Burhan, Muhammad, et al. "IoT elements, layered architectures and security issues: a comprehensive survey." *Sensors* 18 (9) (2018): 2796.

[3] Razzaq, Mirza Abdur, et al. "Security issues in the Internet of Things (IoT): a comprehensive study." *International Journal of Advanced Computer Science and Applications* 8 (6) (2017): 383.

[4] Aqeel-ur-Rehman, Sadiq Ur Rehman, et al. "Security and privacy issues in IoT." *International Journal of Communication Networks and Information Security (IJCNIS)* 8 (3) (2016): 147–157.

[5] Jing, Qi, et al. "Security of the Internet of Things: perspectives and challenges." *Wireless Networks* 20 (8) (2014): 2481–2501.

[6] Hassan, Wan Haslina. "Current research on Internet of Things (IoT) security: a survey." *Computer Networks* 148 (2019): 283–294.

[7] Fazion, Mauro. "Vulnerabilities and security issues of IoT devices." *Sikur Report* 01022020 (2020).

[8] Hossain, Md Mahmud, Maziar Fotouhi, and Ragib Hasan. "Towards an analysis of security issues, challenges, and open problems in the internet of things." *2015 IEEE World Congress on Services*, 2015.

[9] Anthi, Eirini, et al. "A supervised intrusion detection system for smart home IoT devices." *IEEE Internet of Things Journal* 6 (5) (2019): 9042–9053.

[10] Benkhelifa, Elhadj, Thomas Welsh, and Walaa Hamouda. "A critical review of practices and challenges in intrusion detection systems for IoT: toward universal and resilient systems." *IEEE Communications Surveys & Tutorials* 20 (4) (2018): 3496–3509.

6 HomeTec
Energy Efficiency in Smart Home

Ashish Sharma, Sandeep Tayal, Parth Rustagi, Priyanshu Sinha and Rohit Sroa

CONTENTS

6.1 Introduction 95
    6.1.1 Smart LEDs 96
    6.1.2 Smart Shower 96
    6.1.3 Thermal Cooler 96
    6.1.4 Auto Power Cut 96
    6.1.5 Smart Solar Panels 96
6.2 Smart LED 96
6.3 Smart Shower 98
6.4 Thermal Cooler 99
6.5 Auto Power Cut 100
6.6 Smart Solar Panels 101
6.7 Conclusion 102
References 104

6.1 INTRODUCTION

The worldwide energy demand is rising constantly. While many sectors have been trying to reduce their energy consumption for several years, sustainability in the residential domain must still be considered being in its infancy. Realizing an energy efficient building operation is closely tied to the employment of building an automation system, which is considered as an almost mandatory condition for sustainable home. A novel approach to realize the smart, minimum energy, green building is taken in this work. The proposed home system concept is called HomeTec. HomeTec aims at the realization of an intelligent home by introducing semantic context, machine learning and Artificial Intelligence. Our system covers five major equipment and appliances of a smart home, which are LEDs, cooler/thermal control, solar panels, auto cut and smart shower control. All these parts are briefly discussed in the succeeding points.

DOI: 10.1201/9781003097198-6

6.1.1 Smart LEDs

Smart LEDs are self-controlled with the help of Artificial Intelligence and Machine Learning. These LEDs will dynamically strike balance between the natural lightning and powered lightning using smart sensors.

6.1.2 Smart Shower

It will help boost water saving all over the world. This device will allow you to set an efficient timer for each bath and after that will automatically cut the water supply.

6.1.3 Thermal Cooler

With the help of smart sensors all the around, we will be able to keep track of the outside temperature and other weather details like humidity. With the help of thermal cooler our system will help guide the thermal and cooling devices to maintain a comfortable temperature for the user and will guide to shut down when not needed.

6.1.4 Auto Power Cut

It is a part of our system that analyzes those appliances that have not been used for a particular time and then auto cuts power to them. Like in a modular kitchen all the time appliances are connected to power even when not in use similar to TV and gaming consoles. It will be equipped with modifiable and manual exception as defined by the user.

6.1.5 Smart Solar Panels

These panels will provide direct supply of electricity to all the devices in the house. For the night, it will store energy corresponding to the storage units.

6.2 SMART LED

As the computer is getting much powerful and smaller in size, its involvement in various other products is also increasing. Smart LED is one of the products that have seen a major change in its technology by this amalgamation. In simple terms, smart LEDs are just LED bulbs with a touch of intelligence.

It has a processor and control system of its own that makes it smarter than its counterparts (i.e., incandescent bulbs, halogen bulbs). What's special about these LEDs is that they are connected to the internet (or a hub in some cases). This allows the user to have full control over the operation of the bulb from anywhere, that too, with just a simple app on your smartphone, voice control or virtual assistants like Alexa, Siri or Google. The user can change the color of the light from the bulb and brightness of the bulb.

However, smart LEDs are so much more than just changing colors or brightness. Smart LEDs are much more energy efficient than standard LED, fluorescent or halogen

HomeTec: Energy Efficiency in Smart Home

bulbs. To see in this perspective, for a bulb of 80 Lumen, normal LED is 50% efficient than halogen bulbs and 10% efficient than fluorescent bulbs. Further, because of the intelligence in smart LEDs, they offer much more efficiency of about 55% and 15% in comparison to halogen and fluorescent bulbs, respectively. The increased efficiency comes from timer and auto cut features of smart LEDs.

The timer setting of smart LEDs allows you to set a particular time for the lights to go off any time of the day. Geofencing is a technology that uses user's locations to determine whether to switch on/off the lights. Geofencing sets a virtual radius of operation around the house. So if the user leaves his house, all the lights are automatically switched off. Smart LEDs also have sensors that sense the presence or absence of a person and switch on or off the lights, respectively.

Smart LEDs help you set the right mood and lighting according to the occasion, thus omitting the use of party or decorative lighting. It has themes for all festive occasions like Diwali or Christmas. In some cases, smart LEDs help you sleep better. The use of the C Sleep setting in LEDs helps to emit colors of different temperatures to manage the flow of natural melatonin (sleep-regulating hormone) in the body.

It helps to control melatonin levels during the day and increase melatonin when it's time to sleep. It also incorporates lighting science, which emits pleasant light that does not disturb one's sleep cycle.

The cost of a smart LED is on the expensive side. However, it pays back and saves you a whole lot more in the long run. A standard smart LED costs around Rs. 1000. A normal bulb of 80 Lumen has a power output of 60 W [1], whereas that of a smart LED is 2 W. For a normal bulb that runs for 8 hours a day for a month would cost around Rs. 90, whereas for the same condition a smart LED would cost around Rs. 3, thus saving Rs. 87. The saving in a year would be Rs. 1044, which is greater than the cost of smart LEDs. Further smart LEDs are much durable and are rated for 20 years. All of this can be achieved from a single bulb. If all of the bulbs are replaced from households, it would not only save a lot of money but also help reduce electricity consumption, thus promoting a greener and efficient environment (Figure 6.1).

FIGURE 6.1 Smart LED bulb.

TABLE 6.1
Brightness vs Output of Different Bulbs

Brightness(Lumens)->	250+	450+	800+	1100+	1600+
Standard	25 W	40 W	60 W	75 W	100 W
Halogen	18 W	29 W	43 W	53 W	72 W
CFL	6 W	10 W	13 W	18 W	23 W
Smart LED	4 W	5 W	10 W	15 W	20 W

From Table 6.1, we infer that smart LEDs have much lower power consumption compared to other bulbs. Especially at higher lumens, power consumption is very low, thus saving us more power and money.

6.3 SMART SHOWER

It is estimated that by 2030, we will be void of all the natural freshwater resources. Many countries like South Africa and Brazil are already running out of water [1]. Thus, saving water is a high priority and there is no better way to start so from households and moderating the daily activities that involve water. Bathing is one of the activities that involve most quantities of water, in numbers about 17% of total water usage. Out of this about 17%, about 20% of water is wasted daily, which amounts to 2.5 gallons per minute [2]. All of this is from one home. Considering the population, the amount of water that is wasted increases exponentially.

Smart showers technology is a new but effective technology that incorporates computers and Internet of Things (IOT) to save water and energy and provide greater comfort. Smart showers are much like standard showers but with an addition of intelligence. They smartly manage the factors affecting a shower and implement a method that saves water by stopping the supply when not in use. The factors affecting a shower like the temperature of water or time of a shower can be controlled using a digital dial on the shower or an app on a smartphone.

By setting specific times for shower, the shower can smartly manage the energy used to heat the water for only the required time and not heating the water all the time. This saves electricity. Some showers use atomized nozzle which reduces the size of water droplets and makes the flow faster. Further, the temperature of the water can also be regulated by a nozzle with little to no electricity and cover twice the area than a standard shower. Smart showers can also limit the time a person spends in a shower. In case of no activity, it would switch off all of its operations and sit on idle, which saves the cost of unnecessary heating and/or flow of water.

The use of smart showers has saved 100 million gallons of water in the previous years. A typical shower heater uses 1.6 kWh of electricity to heat the water to hot shower temperature [1], whereas smart showers use little to no electricity to heat water. A smart shower is expensive at first, costing around Rs. 10,000. However, the amount saved in electricity and water bills can compensate for the cost of the shower in a year (Figure 6.2).

HomeTec: Energy Efficiency in Smart Home

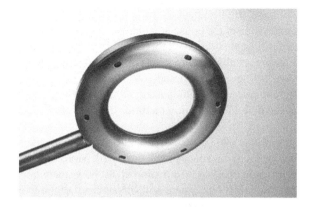

FIGURE 6.2 Smart shower head with atomized nozzle.

TABLE 6.2
Comparison Table for Standard Shower and Smart Shower

Parameters	Standard Shower	Smart Shower
Water consumption	65–88 liters	55–65 liters
Power consumption (hot shower)	1–8 kWh	0–1 kWh (uses atomized nozzle)
Cost	Rs. 10,000–22,000	Rs. 30,000–50,000

From Table 6.2, we infer that smart showers save much more energy and water as compared to standard showers. The increased installation cost of smart showers can be compensated by the savings in energy and amount of water.

6.4 THERMAL COOLER

According to the reports, it has been found that 57% of the residential energy consumption is from heating or cooling appliances. This energy consumption can be reduced to a significant amount using smart home devices. There can be a case in which in winter on a sunny day the shutters can be opened which would let the sunlight to traverse through windows and transparent doors which would ultimately result in heat gain.

Energy consumption can be further reduced by knowing the initial temperature of different parts of the house and determining which areas differ in temperature. Now, apart from artificial appliances this problem can be solved using the natural air flow, where outside air can be used in those areas which differ. But obviously this solution will require to have great knowledge of thickness and material of the exterior as well as interior walls. And because of different construction ways, this can create a very complex problem to understand, whereas our system HomeTec can solve this problem.

Our system will have an intelligent control system considering the knowledge of weather data and the design of the house. It will first scan the whole house and measure the temperatures of different areas. With this initial knowledge and the weather forecast data, it will automatically predict how much temperature would be needed in time. It will be compatible and integrated with different devices manufactured and processed by different companies. Our system would not only keep the temperature balanced but with the help of ventilation facilities it would also assure the air quality inside the house.

The outdoor conditions can be measured and obtained with the help of rain and humidity sensors which will help our system work well. More sensors can be installed in the house which would provide information about how many persons are present in an area according to which our system will keep control of the temperature. It will also keep a measure of air quality in order to make sure the users feel comfortable and keep the share of CO_2 in the air at a healthy level.

With smart connections with shutters and blinds, our system will be able to make choices whether to ventilate natural air or to use the thermal cooler for air conditioning. Like, for example, if the outside temperature is optimal it would open the blinds or shutters to let the natural air flow in and in case if the sensors senses high wind that is not optimal for the interior temperature 98 wood close the shutters and turn on the thermal or air conditioner (Figure 6.3) (Table 6.3).

6.5 AUTO POWER CUT

An auto power cut software is capable of immediately sensing those appliances which have not been in use for a long time or it can also be used for emergency situations or in cars to help prevent an accident. If the vehicle is running at a high speed, an automatic power cut off can prevent the vehicle from exploding, hence saving lives of passengers. This device consists of a current detection circuit, a main control circuit, a host socket and an output control for the peripheral socket power.

FIGURE 6.3 Depiction of thermal cooler and heater controlled using smartphone.

TABLE 6.3
Comparison Table for Different Seasons and Their Energy Savings

Season	Device Type	Efficiency (Low/High)
Summer	Air Conditioner	Low
	Heater	High
Spring	Air Conditioner	Moderate
	Heater	Moderate
Winter	Air Conditioner	High
	Heater	Low

An automatic power cut off timer circuit switches OFF a load beyond a certain fixed period of time. This circuit can be used for home appliances like TV, gaming consoles, kitchen appliances, computer and its peripheral devices. The most common circuit used for this purpose consists of a nonstable multi-vibrator using a 555 IC (Integrated Circuit) [2]. This circuit provides the output for a single triggering pulse. When power is provided to the circuit, it remains on the ON position and triggers itself for time duration to switch off present [3].

Software can be designed for this purpose which can control the home appliances; power supply optimization can vary from appliance to appliance depending upon its usage and also the time it has been on state. Power OFF circuit can help in turning OFF, sleep mode or can restart the appliance at a specific time. The system tray will be controlled by the power off timer circuit whenever the time is assigned to the circuit and will work silently in the background. If any change in the device is observed like a button is pressed, then the device will be in its original state and will perform the task as usual.

This concept can also help in device efficiency and to run that device for long run. Internal health checks for the appliance should also be kept in mind. This will prevent extra load on the device and depending upon its surrounding it can optimize itself. In some countries, power outage is very common; this can damage the device for long run due to sudden fluctuation of receiving power. So we can optimize in such a way that it can delay the power before providing it to the appliance.

If the consumer doesn't want to power cut a device or he wants to use it all the time, solar power can also contribute in optimizing the electricity by supplying direct power to the device.

6.6 SMART SOLAR PANELS

Smart solar panels are the future of conserving electrical energy, which will perhaps promote the efficient scalability of distributed energy resources (DER). Smart solar panels proceed one step ahead to store the solar energy for night time and using the direct supply of energy for products during day time. It can also support functionalities like regulation of voltage and support of frequency. As the number of DER devices increase, the number of grids also increases in solar panels. That's why additional support for smart solar panel functionality has increased.

A smart home with solar inverting functionality must provide sufficient electric power to fulfill all the power requirements of home devices such as lighting systems, microwave, refrigerator, entertainment systems and computers.

Solar home demands requisite number of solar panels to absorb as much as viable solar energy possible. When revealed to direct rays of the sun, a general solar panel home induces about 280–320 Watts in an hour, which can further produce up to 2.5–3.3 kWh in a day [4]. This can differ depending upon the amount of time sunbeams is received during the daytime. This architecture requires a battery that is charged by solar power and can also accumulate electric power to be used in the night. Batteries are usually used for off-grid systems. The arrangement needs an inverter to induce AC power from DC power produced by solar panels so that it can be used by home appliances and other home-oriented devices. The whole structure is integrated with sufficient and requisite cabling and wirings to channelize the energy into usable form.

Solar home systems work on the principle of absorption of sunlight, which strikes the solar window panels, by the photovoltaic cells and the semiconductors of silicon due to the photovoltaic effect, and this solar energy is directly converted into electrical power. This electric power is in the form of DC power which promptly charges the battery.

The main factor which should be taken in account before setting a home solar power plant is to perform some calculations to know the amount of solar panels required for that home. The best way is to use the highest ever paid electricity bill in the recent months or a year. This will help in knowing the daily consumption by comparing the units of electricity used in that month. The solar panel setup must be facing in the direction of south for receiving maximum sunlight. When we know the amount of solar panels required, the vacant space in the house for the setup of solar panels should be kept in mind.

When a solar power system or a solar battery is installed, the system is able to occupy excessive solar electricity in our homes in contrast to sending it back to the grid. If by any chance the solar power is generating more than required energy, this excess energy can be moved toward charging the battery. During night, solar panels don't induce electricity; then we can extract energy that we occupied earlier. In short, the electricity will be sent back to the grid if battery is fully charged and the electricity can only be withdrawn when the battery is exhausted. This basically means that in practical words the home with smart solar panel technology can store excess solar power for later use when there is sunset. As an advantage, smart solar panels can also provide short-term backup if there is a power outage in that area (Figure 6.4) (Table 6.4).

6.7 CONCLUSION

This chapter presented a comprehensive approach to fully unleash the energy saving potential of smart homes. With the knowledge of previously unconsidered information, we have built our system. It integrates conventional data in a unified way and also from a multitude of new parameters come from the architecture, engineering and construction domain like material, thermal property, building layout and orientation.

HomeTec: Energy Efficiency in Smart Home

FIGURE 6.4 Polycrystalline solar panel powering a household.

TABLE 6.4
Comparison Table for First Generation Solar Panels

Type of Solar Panel	Efficiency	Watt Range (W)	Cost of Panel (₹/ per W)
Monocrystalline solar panels (MonoSI)- These comprise monocrystalline silicon. They are mostly costly due to their nature of consuming less space, high force yield and long strength.	17%	250–above 300	₹ 47
	18%	250–300	₹ 48
		Above 300	₹ 50
	19%	0–50	₹ 46
		200–300	₹ 42
Polycrystalline solar panels (Poly-SI)- Their creation innovation depends on softening crude silicon. Their external structure has square cells. They are less expensive than mono-SI since they consume more space to produce a similar measure of vitality in comparison to mono-SI, have lower proficiency and shorter life expectancy and cannot endure amazingly hot temperatures.	13%	0–50	₹ 64
		150–200	₹ 52
	14%	0–50	₹ 88
		200–250	₹ 52
	15%	50–100	₹ 63
		250–300	₹ 37
	16%	50–100	₹ 68
		Above 300	₹ 37
	17%	0–50	₹ 73
		Above 300	₹ 36

We introduced a first part of the system intelligence, namely knowledge inference and reasoning, at a very early design stage. It allowed us to make some decisions already on the data level, thus facilitating the higher control tasks.

REFERENCES

[1] Wolfgang Kastner, Félix Iglesias, MarioJ Kofler, and Christian Reinisch, "ThinkHome energy efficiency in future smart homes," *EURASIP Journal on Embedded Systems* 2011, p. 18, 2011.

[2] Myung Soon Bae, Daeback Apt, Okkye Dong, Gumi shi, and Kyungsangbuk do, "Automatic power cut-off device for emergency situations," 6, 111,327, March 1, 1999.

[3] Raneen Ayman Alzafarani and Ghadi Ahmad Alyahya, "Energy efficient IoT home monitoring and automation system," *2018 15th Learning and Technology Conference (L&T)*, p. 5, February 2018.

[4] Ana Koren and Dina Šimunić, "Modelling an energy-efficient ZigBee (IEEE 802.15.4) body area network in IoT-based smart homes," *2018 41st International Convention on Information and Communication Technology, Electronics and Microelectronics (MIPRO)*, May 2018, p. 5.

7 Impact and Suitability of Reactive Routing Protocols, Energy-Efficient and AI Techniques on QoS Parameters of WANETs

Meena Rao and Richa Gupta

CONTENTS

7.1 Wireless Ad Hoc Networks 105
7.2 Types of Routing Protocols 106
    7.2.1 Classification of Routing Protocols 106
        7.2.1.1 Proactive or Table-Driven Routing Protocols 106
        7.2.1.2 Reactive or On-Demand Routing Protocols 107
7.3 Summary 116
References 116

7.1 WIRELESS AD HOC NETWORKS

Wireless ad hoc networks (WANETs) are built to connect two or more than two devices with each other without any underlying infrastructure. There are no separate routers or modems for the transmission of data. The nodes themselves act as routers, finding routes and transmitting data packets to the next node or to the destination node. A WANET is a collection of mobile nodes with no pre-established or fixed architecture [1, 2]. In this type of network, nodes act as routers by relaying each other's packets and all the nodes form their own cooperative infrastructure [3]. The nodes can communicate through single hop or multihop paths. Due to the characteristics of WANETs, they can be used to establish connectivity in areas where network needs to be established quickly (emergency medical situations, military operations), without much overall cost involved and without any backup support present. Although these features of WANETs lead to an easy setup of network, such systems have inherent

DOI: 10.1201/9781003097198-7

issues as well. The major issues with these networks are dynamic topologies, bandwidth constraints, variable capacity links, energy-constrained operations and complete self-organized behavior. Hence, to have a stable and robust network that allows proper connectivity and suitable transfer of information, the network nodes of WANETs should be supported by efficient routing protocols. To provide a reliable setup that adheres to certain Quality of Service (QoS) parameters, it is necessary to ensure that an optimum route is found between source and destination but due to the dynamic nature of MANETs, the routing problem is much more complicated as compared to wired networks [4]. Besides finding an optimum route, it is also imminent to provide various QoS parameters like good throughput, minimum delay, least packet loss, less jitter [5]. Many multimedia and real-time applications like file sharing, video conferencing, mobile learning require high bandwidth and have stringent delay, jitter as well as packet loss requirements. Providing real-time or multimedia applications with QoS guarantees is quite a challenging task as these applications demand high bandwidth and are delay-sensitive in nature.

Moreover, the inherent nature of MANETs is characterized by frequent link breakages and node failure due to which providing QoS in such networks still becomes more difficult [6, 7].

7.2 TYPES OF ROUTING PROTOCOLS

Wireless as well as ad hoc networks mainly use on-demand-based routing protocol. The routing protocol in MANETs should have various characteristics like they should be adaptive to frequent topology changes. Also, the number of broadcast packets should be kept minimum as it will lead to packet collisions. Reliable transmission should be implemented using effective routing techniques to reduce message loss. Moreover, routing should be fully distributed in nature since centralized routing involves high control overhead and also there is a chance of single point failure. To reduce control overhead, routing is generally localized. Also, the nodes are generally small, portable, hand-held devices with limited battery availability and energy constraints. Hence, bandwidth, computing power, memory and battery should be judiciously used in these networks. Every node in a WANET should try to store information regarding the stable local topology only. Frequent changes in the local topology and changes in the remote parts of the network need not be updated in the topology information maintained by the node. In efficient routing protocols, routing overhead is avoided and this is quite important in a bandwidth-constrained environment.

7.2.1 Classification of Routing Protocols

7.2.1.1 Proactive or Table-Driven Routing Protocols

Here, every node maintains the network topology information in the form of routing tables [8]. To find a path from source to destination, the node runs an appropriate path finding algorithm. This type of protocol immediately provides the required routes when needed but at the cost of bandwidth which is used in periodic updates

Impact of Reactive Routing Protocols

of topology. Proactive protocols can be said to be an extension of the wired routing protocols. Proactive protocols or table-driven protocols also utilize a large amount of bandwidth especially in systems like WANETs which already have strict bandwidth constraints. Moreover, they result in the formation of long and short loops. In these protocols, the complete routing information is maintained in the form of tables at every node. Since the node positions change frequently especially in a dynamic system, the tables need to be updated frequently to maintain the correct routing information. This results in unnecessary routing overhead and more bandwidth is required for constantly maintaining routing table information. The overhead and bandwidth consumption becomes all the more high when nodes are dynamic. Some of the important proactive routing protocols are Destination Sequenced Distance Vector (DSDV) and Optimized Link State Routing (OLSR).

7.2.1.2 Reactive or On-Demand Routing Protocols

Reactive routing protocols do not maintain any network topology information and were designed to reduce the overheads in proactive protocols by maintaining information for active routes only. Reactive protocols save a lot of control overhead as there is no need to exchange routing information periodically. The necessary route from source to destination is acquired as and when required through a connection establishment process. Routes are usually discovered by flooding route request (RREQ) packets in the network. On receiving an RREQ packet, a route reply (RREP) is sent back to the source. In case the destination is directly reached, reply is sent back via a bi-directional link or otherwise it is sent through piggybacking [9]. Reactive routing protocols are further categorized as source routing and hop-by-hop routing. In source routing, the data packets carry the complete source to destination address. The advantage of such a method is that intermediate nodes need not maintain updated routing information. The intermediate nodes also do not need to have an idea about the neighboring nodes. Though this technique reduces overhead, it does not work efficiently in large networks. In large and dynamic systems, during the process of sending packets from source to destination, some intermediate nodes may be added in the network or some nodes may change their position resulting in the change of path of original source to destination. This may result in packet loss, error and hence poor QoS parameters.

The second method is hop-by-hop routing or point-to-point routing. Here, each data packet instead of carrying the entire route information carries only the address of the next hop. Every node has a routing table and the packets are forwarded from every hop to the destination with the help of this routing table. One advantage of this method is that better and shorter routes can be selected as routing tables are updated periodically to reflect any changes in the topology. This method is especially beneficial in a dynamic system. However, every node needs to store information of every active node in its routing table and also the information of the neighboring nodes has to be saved. This becomes cumbersome and disadvantageous in a large and dynamic network.

Some of the common reactive routing protocols are as follows:

Dynamic Source Routing Protocols:

In DSR protocol, a route setup message makes a record of all the nodes it has passed through and based on this record an optimal data exchange path is selected by the destination node. It is an on-demand protocol that minimizes the bandwidth consumed by control packets [10]. Minimization of bandwidth is mainly because periodic HELLO packets (beacon messages) that are used by the node to inform the neighbors of its presence are not required. Also, intermediate nodes do not require to store any routing information. A route is established by flooding RREQ packets in the network and the destination node sends the RREP packets back to the source. Each node on receiving the RREQ packets forwards or broadcasts the packet to its neighbors if it has not been forwarded already. If the node that receives the RREQ packet is itself the destination node, then the packet is not further broadcasted and an RREP packet is sent. To avoid multiple transmissions, the sequence number on the RREQ packet is checked by an intermediate node before it is forwarded further. This avoids duplicity or loop formations of the same RREQ packets.

Several optimization techniques have been further introduced to improvise the performance of DSR protocol. It uses route cache at intermediate nodes. Intermediate nodes use this cache information to reply to the source on receiving an RREQ packet. Route cache is also used by intermediate nodes to provide a route to the corresponding destination. Piggybacking of data packet can also be done so that RREP packets can be sent along with RREQ packets.

Moreover, if optimization is not allowed in the DSR protocol, then route formation becomes simpler. In Figure 7.1, suppose node 1 is the source node and node 11 is the destination node. Node 3 receives an RREQ packet from suppose both node 2 and node 4. Node 4 discards one of the RREQs and considers the other RREQ and

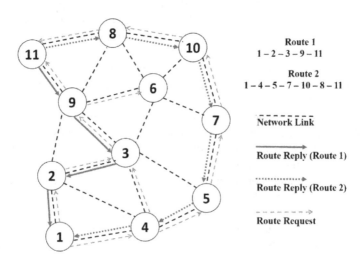

FIGURE 7.1 Working of DSR protocol is presented with route reply and route request message being sent between various nodes.

forwards the packet through the intermediate nodes to the destination. So, one of the routes can be 1-2-3-9-11.

The advantage of DSR protocol is that a route is established only when it is required and the constraint of finding the routes for all other nodes is avoided. The route cache information is utilized by intermediate nodes to find routes to the destination. However, maintenance of route cache information becomes extremely difficult in a dynamic system like WANETs and it is also difficult to maintain route cache in a large network. Also, the DSR protocol does not provide any solution for route or link breakage. This means that if any intermediate node between source and destination fails, it leads to packet loss and low QoS parameters. Any solution like provision of backup routes or route repair mechanism is not provided.

- Split Multipath Routing (SMR) Protocol:

SMR is an on-demand routing protocol. SMR builds multiple routes and this is done using a request/reply cycle [11]. When a message has to be sent to the destination and route information is not known, the RREQ message is flooded in the entire network. Since RREQ is flooded throughout the network, multiple messages reach the destination through different routes. SMR protocol tries to build multiple paths to the destination to prevent congestion at nodes. However, to achieve this, all available routes to the destination must be known which is very difficult in a dynamic system. This would also involve lot of control overhead which is again not desirable in a bandwidth-constrained environment. For data packet transmission, source transmits an RREQ packet with a source identifier (ID) and a sequence number. If any node that is not a destination node receives the RREQ, it checks RREQ for duplicity. If it is a fresh RREQ, then its identifier is appropriately updated and packet is further retransmitted. In SMR, intermediate nodes do not send replies even if route to the destination is known. Also, the destination may receive multiple RREQs and for these multiple RREQs, destination sends RREPs for each individual route. However, SMR generates multiple paths that are overlapped and also multiple RREPs which may lead to flooding in the network. SMR protocol also does not have any provision for route repair or provides a backup route. Flooding of the network and lack of any route repair mechanism make SMR protocol not quite suitable for WANETs as it reduces the QoS parameters.

- Ad hoc On-Demand Distance Vector (AODV) Protocol:

AODV uses an on-demand routing approach for finding routes. It closely adapts the DSDV protocol in ad hoc wireless networks. In AODV, the route is established only when there is a need for data transfer at the source node. Hence, AODV is also an on-demand scheme. AODV also employs a destination sequence number to identify the most recent path [12]. In DSR, source routing is used in which a data packet carries the complete path address. However, in AODV, source node and intermediate nodes store the next hop information. Similar to other on-demand routing protocols, an RREQ is flooded in the network. However, AODV uses a destination

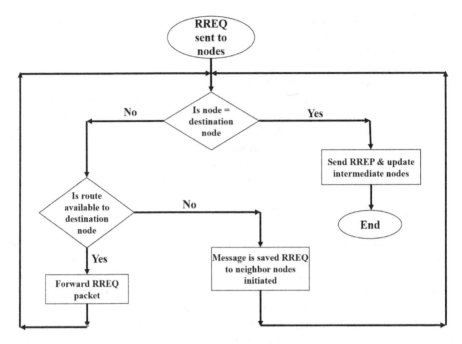

FIGURE 7.2 Flowchart explains AODV Protocol and how RREQ is sent to destination nodes.

sequence number (DestSeqNum) for obtaining an updated route to the destination. RREQ packets are broadcasted to the neighboring nodes. An RREQ carries the source identifier (SrcID), the destination identifier (DestID), the source sequence number (SrcSeqNum), the destination sequence number (DestSeqNum), the broadcast identifier (BcastID) and time to live (TTL). The neighboring nodes broadcast the RREQ to their neighbors. On receiving an RREQ from an intermediate node, an RREP is sent to the source node if the intermediate node itself happens to be the destination node itself or else RREQ is forwarded to the next node. Multiple RREQs received at a node are discarded. Multiplicity of RREQ packets can be determined by the BcastID-SrcID pair. RREP from the most recently updated node is sent back to the source node. Route entries or entries of the previous nodes are made when an RREQ is forwarded by an intermediate node. This is helpful when the destination node is reached as the RREP can be sent via all the updated addresses that have been saved. Obsolete entries, which are not used within a specified time, are deleted. Figure 7.2 shows a simple flowchart of AODV protocol implementation. It shows that RREQ packets are sent to the network nodes.

A node is first checked to verify if it is the destination node. If it is the destination node, RREP is sent to the source or else it is checked if a route to the destination node is available. If a route to the destination node is available, the RREQ packet is forwarded or else the message is saved and RREQ to the neighbor nodes is initiated. The inclusion of localized flooding during routing leads to a better packet delivery ratio. Also, the use of DestSeqNum helps to establish the latest route. AODV was further improvised by including a hop count in routing table and data packets. However,

Impact of Reactive Routing Protocols

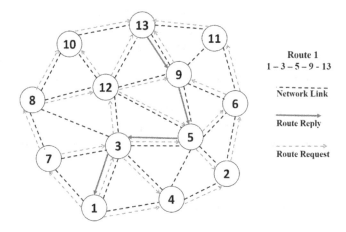

FIGURE 7.3 Working of AODV Protocol is presented with route reply and route request messages being sent between various nodes.

AODV does not undertake route repair. So in case of link breakage, an alternate route establishment process takes place which leads to packet loss as well as delay which is undesirable. Moreover, it is not possible to maintain a hop count when the systems become large and dynamic. This results in a scalability problem. So, AODV protocol also needs to be modified suitably with the help of optimization and other techniques so as to solve the routing problems of dynamic and large systems.

In Figure 7.3, suppose node 1 is the source node and node 13 is the destination node. To find a route to the destination, an RREQ is flooded in the network. Nodes 4, 3 and 7 on receiving the RREQ packet try to find a route to the destination. On availability of a route, RREQ is further forwarded to the appropriate nodes. Since nodes 2, 5 and 8 are the neighbors of nodes 4, 3 and 7, respectively, and if these nodes also have the route to the destination, then a path 1-3-8-10-13 or a path 1-3-5-9-13 to the destination may be established. The RREQ may reach the destination node via different paths and multiple RREPs are sent to the source. All intermediate nodes update their routing tables accordingly.

AODV protocol is quite widely used in wireless networks. There is no flooding in the network and route replies are also received. Thus, QoS parameters obtained with AODV is also found to be better than DSR protocol. However, when using in emergency medical situations, critical business meetings and other important scenarios, it becomes necessary to have backup routes so as to ensure the smooth delivery of packets in the case of route failure.

- Ad hoc On-Demand Multipath Distance Vector (AOMDV) Protocol:

AOMDV is designed to provide efficient and fast recovery from route failures [13]. AOMDV extends AODV to provide multiple paths during route discovery. This protocol uses a hop-by-hop approach and establishes multiple reverse paths from the destination node as well as various intermediate nodes. The basic idea behind developing this protocol was to create multiple paths from source to destination that are loop-free. Frequent route updates take place at all nodes for avoiding loop formation. The routing information available in the underlying AODV protocol is utilized to reduce overhead and provide loop-free routing. Packet delay time is reduced significantly when using AOMDV protocol, often by more than a factor of 2. Presence of multiple paths leads to less packet loss when route breaks occur. The AODMV protocol still needs to be tested for scalability and for heavy traffic. A major disadvantage of this protocol in a resource-constrained environment is that it may lead to flooding of the network as same packets may go through several paths to reach the destination. This results in packet duplicity and is certainly not desirable when resources are limited as in WANETs.

- Ad hoc On-Demand Distance Vector Backup Routing (AODV BR)

The traditional AODV BR relies on creating a mesh-like structure to provide an alternate route. The neighboring nodes overhear the packets that are being transmitted and this way establish an alternate path. During the RREP phase, nodes hear many RREP packets for the same destination, it chooses the best among them and that route is accommodated in the routing table. Although AODV BR shows better performance in terms of total number of processed messages, it results in many drawbacks also. The major drawback associated with AODV BR is that on failure of the backup route, there is no way packets can be delivered to the destination leading to loss and error in the network. Another reactive routing protocol Dynamic Backup Routes Routing Protocol (DBR^{2}P) reconstructed several routes from source to destination so that on failure of the main route, other routes are available for packets to reach the destination. Presence of multiple routes results in less packet loss as apart from the main route, there are other paths through which data packets can reach the destination. However, multiple paths leading to the same destination may lead to flooding of the data packets in the network and also high bandwidth consumption. SMR protocol also tries to address the problem of link failure in a similar way by reconstructing several routes from source to destination. Reconstructing several routes may also lead to flooding in the network which is not desirable. Also, the presence of multiple routes at the same time results in the consumption of a lot of bandwidth and more control overhead. This is not desirable in bandwidth-constrained systems like MANETs. Also, the backup routes are not selected on any specific criteria wherein the best possible backup route is selected. Moreover, the backup routing protocol was not found to work well in high mobility scenarios. SMORT (Scalable Multipath On-Demand Routing for Mobile Ad hoc Networks) provided intermediate nodes on the main path with multiple routes in case the primary path fails. This protocol worked well in high mobility and increased traffic conditions but routing overhead and flooding in the network due to the presence of multiple nodes remained a concern. Further solution was provided by providing a backup route along with a primary route to create a backup path in case of link failure. Here, a primary and a secondary backup path are created as a result of a route control message exchange process. Control messages contain information for guaranteeing service quality. After detecting a failure when sending data, a repairing procedure occurs near the failed node of the primary path. Here again, the issue of backup route failure is not addressed.

In the backup route method, multiple alternate routes are created and maintained along with the main route in a fish bone structure [14]. This technique provides more options of routes to the destination as compared to AODV protocol. This results in successful delivery of packets even in the case of route failure. However, creating and maintaining multiple routes result in more control overhead and extra efforts for maintenance which is not desirable. Creating and maintaining routes are all the more difficult when the system is large and dynamic as positions of the nodes keep changing again and again. When a node failure occurs in the AODV BR scheme, then there will exist several nodes that can transmit several copies of the data packet to the destination node. In a bandwidth-constrained environment, this results in resource wastage. Moreover, overhearing of the RREP packets by neighboring nodes in addition to keeping multiple copies of the data packets compromises security and also requires more processing power and storage. Another technique that was proposed in backup routing was to repair routes locally [15]. These route repair procedure takes place near the broken link so that time consumed in route rediscovery is minimized. However, the route repair procedure also takes time and if route repair cannot be done or takes more than time to live of the packet, it may result in packet drop or packet loss which is not desirable at all. Two other route repair techniques Ad hoc On-Demand Distance Vector Local Repair (AODV LR) and Ad hoc On-Demand Distance Vector Adaptive Backup Routing (AODV ABR) have also been used by researchers. AODV LR is used to repair the route when broken routes are near the destination. Here, routes can be repaired and packets are delivered fast as route repair takes place near the destination. However, AODV ABR repairs route along the primary path when alternate routes exist. AODV LR and AODV ABR do not provide any backup routes to support packet delivery and improvement of QoS.

Backup routing protocols especially AODV BR are found to perform better in terms of various QoS parameters as compared to the traditional AODV and DSR protocols. It is seen that due to low error rate and better packet delivery ratio, AODV BR can be implemented in ad hoc networks in critical scenarios. However, there are certain critical situations (medical or military) where packet drop is not at all a desirable and affordable situation. AODV BR protocol although provides the option of a backup route but does not consider the situation and provide the solution when even the backup route fails. AODV nthBR protocol discussed below addresses this issue.

- AODV nthBR Protocol (Energy-Efficient Routing Technique)

AODV nthBR provides multiple backup routes in AODV environment. The backup routes are provided as and when required. Initially, when packets have to be transmitted to the destination, only a single, most efficient route is made available to the destination. Multiple backup routes to the destination do not exist. Since multiple routes to the destination do not exist, duplicity of packets and flooding in the network are avoided. A backup route is created only when the main route fails. Nodes

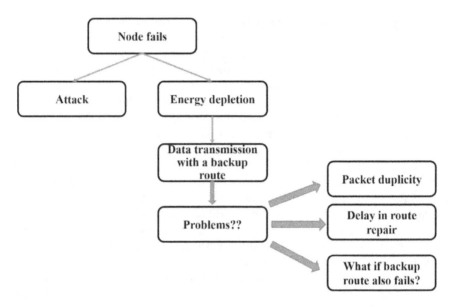

FIGURE 7.4 Depicts node failure due to energy depletion.

for backup routing are selected on two criteria: its distance from the broken route and energy efficiency of the node. When the original route fails, the first backup route comes into picture, when the first backup route also fails, then the next backup route carries out the packet delivery and so on. As long as nodes are available and have energy up to a threshold level required for transmission, packet delivery can be achieved. Main advantage of AODV nthBR protocol is that the selection of nodes for routing is done efficiently on the basis of distance and energy available with the nodes. There is no duplicity in data packets that are transmitted to the destination as data packets are not simultaneously transmitted on multiple routes.

Figure 7.4 describes the problems that occur on node failure and working of AODV nthBR protocol that helps overcome the problem. It is explained that when a node failure occurs, it is either due to an external security attack or depleted energy of the node. Rao et al. [16, 17] consider the case of route failure due to depleted energy of the nodes. When a node is depleted of any energy, the node either becomes dead or inactive and, in turn, incapable of data transmission. This leads to link breakage or route failure. When route failure occurs, it leads to packet loss and unwanted error.

So, a solution in the form of AODV nthBR protocol is presented. AODV nthBR protocol is explained with the help of Figure 7.5 in which nth backup routing is implemented by finding the nearest node to the failed node. After the node is discovered, it is checked for energy efficiency. As can be seen from the decision box of the diagram, if energy efficiency of the node is more than a minimum threshold value, then the particular node is selected for backup routing in data transmission. If energy efficiency is less than the minimum threshold value, then the process of finding the next nearest node for data transmission continues.

Impact of Reactive Routing Protocols

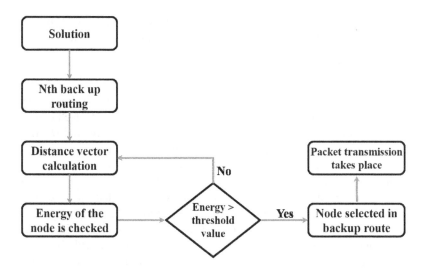

FIGURE 7.5 Working of AODV nthBR protocol is presented along with what happens when node energy is less than or greater than threshold energy.

It is considered that some nodes have α times more energy than the other nodes after the start of data transmission. Hence,

$$Energy\ of\ the\ Node = Initial\ Energy \times (\alpha) \qquad (1)$$

$$Total\ Energy = Initial\ Energy \times (1 + \alpha) \qquad (2)$$

Initially, the dissipated energy is zero and residual energy is the amount of initial energy in a node. Hence, total energy E_t is also the amount of residual energy because it is the sum of dissipated and residual energies. Average distance between the transmitting device and destination D_{bs} is calculated from the following formula:

$$D_{bs} = (one\ dimension\ of\ field) / \sqrt{2\pi k} \qquad (k = 1) \qquad (3)$$

$$D_{bs} = (0.765 \times one\ dimension\ of\ field) / 2 \qquad (4)$$

AODV nthBR protocol is found to perform better in terms of various QoS parameters like throughput, packet delivery ratio and end-to-end delay as compared to the other protocols like AODV, DSR and AODV BR. The node selection is done by energy-efficient criteria; backup routes are available one by one as per requirement for packet transmission. Packets are not flooded in the network. All these factors make AODV nthBR protocol a robust solution to provide efficient QoS parameters. This becomes all the more useful and important when the packet transfer is to be done for critical services. However, there are certain aspects like security of the data that has to be transferred that are not addressed in AODV nthBR protocol.

- Artificial Intelligence Techniques in Routing Protocol

Artificial intelligence (AI) techniques are also being explored by researchers for improvement of QoS parameters in wireless networks. Using AI techniques leads to better collection of data and observation of nodes. Ease of monitoring of nodes leads to better selection of nodes for data transmission resulting in efficient packet delivery ratio and other necessary parameters. Mainly, there are two AI techniques, direct diffusion and energy-aware routing, that are pre-dominantly used [18, 19].

In a direct diffusion technique, a base station broadcasts the information to all the nodes. Nodes receiving the request prepare a gradient toward the requesting node and the process is continued until a direct gradient is set up. The name direct diffusion is because here a targeted node is directly able to communicate with the base node. The direct diffusion technique can be applied when the number of users or nodes is few or in areas with difficult terrains where users are less. Using this technique, information can be collected from the target node and sent to the base station. Another technique that comes under the AI routing protocol is energy-aware routing. Here, the energy consumption of all the nodes is reduced as the information collection as well as packet transfer is done via all the available nodes. Using all the nodes judiciously and in an efficient matter results in a robust technique.

When the number of users is more in a WANET, then a clustering technique can also be used for providing efficient routing protocols. Cluster heads can be selected using self-organizing maps. Cluster management is also an efficient technique and comes into use especially in situations where bandwidth is constrained.

7.3 SUMMARY

In this chapter, impact and suitability of various traditional and reactive routing protocols on WANETs have been discussed. It can be analyzed from the discussion that reactive routing protocols like AODV and its variants like AODMDV, AODV BR and energy-efficient AODV nthBR are suitable to be used for various applications of WANETs. AI techniques and its impact on routing protocols have also been discussed in this chapter.

REFERENCES

[1] Ghalib S., Kasem A., Ali A., "Analytical Study of Wireless Ad-Hoc Networks: Types, Characteristics, Differences, Applications, Protocols", Proceedings of the International Conference on Futuristic Trends in Networks and Computing Technologies, pp. 22–40. 2020.
[2] Hu C.C., Wu E.H.K., "Bandwidth-satisfied Multicast Trees in MANETs", *IEEE Transactions on Mobile Computing*", 7, 6, 712–723, 2008.
[3] Lindeberg M., Kristiansen S., Plagemann T., Goebel V., "Challenges and Techniques for Video Streaming over Mobile Ad Hoc Networks", *Multimedia Systems*, 17, 1, 51–82, 2010.

[4] Gulati M., Kumar K., "A Review of QoS Routing Protocols in MANETs", IEEE International Conference on Computer Communication and Informatics, pp. 1–6, 2013.
[5] Farkas K., Wellnitz O., Dick M., Gu X., Busse M., Effelsberg W., Rebahi Y., Sisalem D., Grigoras D., Stefanidis K., Serpanos D.N., "Real-time Service Provisioning for Mobile and Wireless Networks", *Computer Communications*, 29, 5, 540–550, 2006.
[6] Reddy T.B., Sriram S., Manoj B., Murthy C.S.R., "MuSeQoR: Multi-path Failure-tolerant Security-aware QoS Routing in Ad Hoc Wireless Networks", *Computer Networks*, 50, 9, 1349–1381, 2006.
[7] Jacquet P., Mulethaler P., Qayyum A., Laouiti A., Viennot L., Clausen T., "Optimized Link State Routing Protocol", Internet Draft, Internet Engineering Task Force, 2001.
[8] Ray N.K., Turuk A.K., "Performance Evaluation of Different Wireless Ad Hoc Routing Protocols", *International Journal of Wireless and Mobile Networks*, 4, 2, pp. 203–216, 2012.
[9] Karthikeyan N., Bharathi B., Karthik S., "Performance Analysis of the Impact of Broadcast Mechanisms in AODV, DSR and DSDV", IEEE International Conference on Pattern Recognition, Informatics and Mobile Engineering, pp. 144–151, 2013.
[10] Lee S-J., Gerla M., "Split Multipath Routing with Maximally Disjoint Paths in Ad Hoc Networks", IEEE International Conference on Communications, pp. 3201–3205, 2001.
[11] Perkins C.E., Royer E.M., "Ad Hoc on Demand Distance Vector Routing", IEEE Workshop on Mobile Computing Systems and Applications, pp. 90–100, 1999.
[12] Guizani M., Makhlouf A.M., "SE-AOMDV: Secure and Efficient AOMDV Routing Protocol for Vehicular Communications", *International Journal of Information Security*, 18, 665–676, 2019.
[13] Lee S-J., and Gerla M, "AODV-BR: Backup Routing in Ad Hoc Networks", IEEE Wireless Communications and Networking Conference, Vol. 3, pp. 1311–1316, 2000.
[14] Zhang F., Yang G., "A Stable Back Up Routing Protocol for Wireless Ad Hoc Networks", *Sensors, MDPI Journal*, 20, 23 (6743), 1–11, 2020.
[15] Rao M., Singh N., "An Improved Routing Protocol (AODV nthBR) for Efficient Routing in MANETs", Proceedings of the International Conference on Advanced Computing, Networking and Informatics, pp. 215–223, 2014.
[16] Rao M., Singh N., "Performance Evaluation of AODV nthBR Routing Protocol under Varying Node Density and Node Mobility for MANETs", *Indian Journal of Science and Technology*, 8, 17, 1–9, 2015.
[17] Barbancho J., Leon C., Molina F.J., Barbancho A., "Using Artificial Intelligence in Routing Schemes for Wireless Networks", *Journal of Computer Communications*, 30, 14–15, 2802–2811, 2007.
[18] Lee Z-J., Cho So-T., Lee C-Y., Peng B-Yu., "AODV with Intelligent Priority Flow Scheme for Multi-hop Adhoc Networks", *Vietnam Journal of Computer Science*, 3, 259–264, 2016.
[19] Sendra S., Rego A., Lloret J., Jimenez J. M., Romero O., "Including Artificial Intelligence in a Routing Protocol Using Software Defined Networks", IEEE International Conference On Communications Workshops", Paris, pp. 670–674, 2017.

8 Malicious Use of Machine Learning in Green ICT

Pragya Kuchhal and Ruchi Mittal

CONTENTS

8.1	Introduction to Machine Learning	120
8.2	Current Trends in Machine Learning	121
8.3	Motivation for the Application of Machine Learning in Green ICT	121
8.4	Prominent Machine Learning Algorithms	121
	8.4.1 Supervised Learning	121
	8.4.1.1 Decision Tree	122
	8.4.1.2 Naïve Bayes	122
	8.4.1.3 Support Vector Machine	122
	8.4.2 Unsupervised Learning	122
	8.4.3 Reinforcement Learning	122
	8.4.4 Neural Network Learning	123
	8.4.5 Instance-Based Learning	123
8.5	Introduction to Green ICT	124
8.6	Approaches to Green ICT: Green Internet Technologies	124
8.7	Green Data Center Technologies	125
8.8	Green Communication and Networking	126
8.9	Green Machine to Machine Technology	126
8.10	Green Cloud Computing	127
8.11	Green Wireless Sensor Network	127
8.12	Green RFID Technology	128
8.13	Machine Learning-Based Green ICT	129
8.14	Challenges in Machine Learning-Based Green ICT	130
8.15	Malicious Use of Machine Learning in Green ICT	131
8.16	Solution for Malicious Use of Machine Learning in Green ICT	131
8.17	Applications of Green ICT	132
8.18	Conclusion	133
References		133

DOI: 10.1201/9781003097198-8

8.1 INTRODUCTION TO MACHINE LEARNING

Machine learning (ML) [1–5] is one of the most significant methods inside advancement associations that are searching for inventive approaches to use information resources to help businesses and increase another degree of comprehension. With the appropriate ML models, associations can constantly foresee changes in the research and businesses. Machines are intelligent in nature they can easily learn at whatever point the changes are made like in its structure, program, or information. In any case, for instance, when the performance of a pattern-recognition machine improves after seeing several samples of a person's handwriting, we feel much legitimized all things considered justified/normalized that all the things are considered to state that the machine has learned.

ML [6] for the most part alludes to the adjustments in frameworks that perform undertakings related to Artificial Intelligence (AI) [10, 11]. Such assignments include acknowledgment, conclusion, arranging, robot control, expectation and so on. The changes may be either upgrades to previously performing frameworks or planning of new frameworks. One may ask "For what reason should machines need to learn? Why not configuration machines to proceed as wanted in any case?" There are a few significant building reasons why ML is important [7].

Some of these are:

- Human creators frequently produce machines that do not fill in just as wanted in the situations in which they are utilized. AI strategies [12] can be utilized for hands on progress of existing machine plans.
- The measure of information accessible about specific assignments may be unreasonably enormous for unequivocal encoding by people. Machines that become familiar with this information step by step may have the option to catch a greater amount of it than people would need to record.
- Environments change after some time. Machines that can adjust to a changing situation would decrease the requirement for steady overhaul.
- A few assignments cannot be characterized well, as, we may have the option to indicate inputy matches however not a brief connection among inputs and outputs. We might want machines to have the option to modify their interior structure to deliver the right output for countless example sources of information and in this manner reasonably oblige their information/yield capacity to inexact the relationship understood in the models.
- It is conceivable that there are significant connections between the vast piles of information and relationships. ML strategies can frequently be utilized to extricate these connections (data mining) [13].

ML utilizes many of the diverse algorithms that iteratively gain from information to improve the information and anticipate their corresponding results. As the algorithm train the data/information, it is then conceivable to induce exact models dependent on that information. ML model [8] is the yield/output created when we train the ML algorithm with the dataset. After training the data, when a model is provided with an input, an output will be generated. For instance, a predictive algorithm forms a predictive design. ML has now become vital for creating analytic models [9].

8.2 CURRENT TRENDS IN MACHINE LEARNING

The domain of ML is an adequately youthful that it is growing frequently by designing new formalizations of ML issues driven by handy applications. (A model is the improvement of recommendation frameworks.) A recommendation framework is a ML framework that depends on information that shows connects between a lot of clients (individuals) and a lot of things (items). A connection between a client and an item implies that the client has demonstrated an enthusiasm for the item in some design (maybe by buying that thing in the past). The ML technique recommends different options to a client depending on its interest or its query. ML has the capacity of arranged and portable processing frameworks to assemble and ship tremendous measures of information, a wonder regularly alluded to as "Big Data" [14,15]. ML models additionally give a comprehension on vitality framework usefulness with regard to complex human communications. ML algorithms are used for predicting different scenario in the environment like weather forecasting, wireless sensor networks (WSNs) and smart cities. Practicing the ML algorithms in green ICT includes the exciting new efforts in virtualization, WSNs, data center power utilization and IoT.

8.3 MOTIVATION FOR THE APPLICATION OF MACHINE LEARNING IN GREEN ICT

Over the most recent couple of years with the attention on green computing, and less expensive process and capacity, there has been a flood of enthusiasm for ML in green ICT [15–17]. This is so because green ICT made the revolution in producing the modern energy efficient processors, the density-to-performance ratio has also improved, the expense of saving and overseeing a lot of information has been significantly brought down, able to distribute compute processing across clusters of computers to analyze complex data in record time, and ML algorithms have been made accessible through open-source networks with enormous client bases [18]. These are the reasons which motivated us to use ML under the roof of green ICT.

8.4 PROMINENT MACHINE LEARNING ALGORITHMS

ML is the procedure that naturally improves or learns from the examination or experience, and acts without being explicitly programmed. ML makes the computing processes more productive, scalable and cost-efficient. ML creates the model by analyzing the complex information automatically and precisely. In this section, we explore the following different ML methods [19].

8.4.1 SUPERVISED LEARNING

Supervised learning [20] is one of the basic information handling approaches and needs some external guidance. In this learning machines are trained using well "labelled" training data, and on its basis the machines predict the output. The objective of a supervised learning algorithm is to find a mapping function to map the input variable (i) with the output variable (j). When the training procedure closes, we can find a function from an input i with a best estimation of output j ($f: i \rightarrow j$). The significant obligation of this learning method is to produce a model which represents the relationships and dependency links between input features and figure the outputs. Three acclaimed algorithms under this learning are mentioned below.

8.4.1.1 Decision Tree

This approach solves the classification and regression issues by regularly branching information based on a specific attribute. The decisions lie in the leaves and the information is branched in the hops. In classification tree the decision variable is emphatic/definite (result as Y/N) and in regression tree the decision variable is uniform. Decision tree is appropriate for regression as well as for classification issues. It has high efficiency because of tree traversal algorithm. But sometimes it gets unstable and difficult to control size of tree.

8.4.1.2 Naïve Bayes

It mostly focuses on the content classification. It is used for grouping/clustering and classification. The essential structure of this technique depends upon the contingent likelihood. This makes trees dependent on their likelihood of occurrence. Such trees are otherwise called Bayesian networks [21].

8.4.1.3 Support Vector Machine

It is another most generally utilized strategy used for classification. SVM [22, 23] runs on the paradigm of margin evaluation. It draws margins between the classes. The margins are designed in such a manner that the distance between the margin and the classes is highest, subsequently limiting the classification error.

8.4.2 UNSUPERVISED LEARNING

The Unsupervised Learning Algorithm [24,25] takes in not many highlights from the information. At the point when new information is presented, it utilizes the recently learned highlights to perceive the class of the information. It is primarily utilized for clustering and feature reduction.

8.4.3 REINFORCEMENT LEARNING

Reinforcement learning (RL) [26–28] takes the decision, based on which action is to be implemented such that the ultimate objective is affirmative. The learner has no information which moves to take until it has been given a circumstance. The activity/action made by the learner may impact circumstances and their activities later. RL exclusively relies upon two measures: (i) trial and error search and (ii) delayed result. RL applies to successive dynamic issues in which the learner cooperates with a scenario by continuously taking activities—the outputs—based on its perceptions—its inputs—while fetching feedback regarding each selected activity.

8.4.4 Neural Network Learning

The Neural Network (NN) [29, 30] is attained from the biological idea of neurons. A neuron is a cell-like structure in a brain. To depict the NN, one must see how neuron functions. A neuron has four sections (refer Figure 8.1): (i) dendrites, (ii) nucleus, (iii) soma and (iv) axon. It deals with three layers: (i) input layer takes input (like dendrites), (ii) hidden layer processes the input (like soma and axon), (iii) output layer forwards the determined outcome (like dendrite terminals), as represented in Figure 8.2.

8.4.5 Instance-Based Learning

In [31], the learner studies a specific sort of pattern. It attempts to pertain the similar pattern to the newly inserted information, therefore named as instance based. It is a lazy learner technique that waits for the test information to come and then act on it altogether with the training information.

K-Nearest Neighbor: K-Nearest Neighbor (KNN) Algorithm [32] is a classification scheme. It utilizes a database which is having information key points gathered

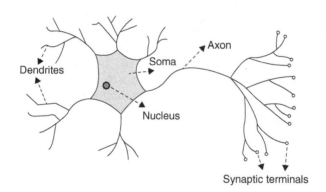

FIGURE 8.1 A neuron.

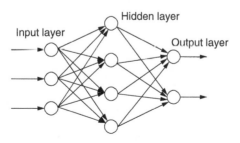

FIGURE 8.2 Artificial neural network's structure.

into a few classes and the method attempts to characterize the sample information point provided to it as a classification issue [33].

8.5 INTRODUCTION TO GREEN ICT

"Green ICT" comprises another term in informatics that is ecologically supportable ICT. The innovative ICT tools, e- and m-administrations and keen advancements in blend with green practices and green conduct either for the ICT new area or the ICT client/resident. This contributes not exclusively to the insurance and reclamation of the earth yet additionally to the improvement of the nature of human life. In this manner, "Green ICT" has instead gotten equivalent to eco-accommodating advancements and programming tools

The green ICT itself comprises several terms as shown in Figure 8.3.

8.6 APPROACHES TO GREEN ICT: GREEN INTERNET TECHNOLOGIES

As of late, green internet is an essential sphere. The internet ideas and technologies contributed are utilized to build up a keen and green network [53]. The usage of vitality in the network gear is obscure on account of the free power. The estimation of the force utilization of network hardware has expressly been considered for estimating exactness and straightforwardness [54]. There exists a considerable capacity to diminish the internet necessities and decrease the unpredictability utilizing capacity of the activity related to planning traffic and switches [55]. Active topology, the board instrument in the green internet, is fabricated and recognized as the shape of cluster and connection structure for vitality utilization in the environment gadget.

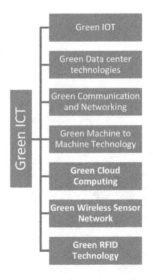

FIGURE 8.3 General terms used in green ICT.

Greening the internet of WAN [39] in the information setup is talked about as the force utilization of assessed wireless access networks [56]. Besides, Suh et al. [57] investigated the impact of the development gear of information networks for the same. The estimation of intensity utilization and sparing vitality capability of information network hardware is a thought. The examination in [58] structures a smart internet steering procedure, so the directing can lead traffic in a greenway. Additionally, the theory is examined through Yang et al. [59] that uncover the separated inexhaustible and non-sustainable power source for green internet steering. Be that as it may, Hoque et al. [60] analyzed method answers to improve the vitality proficiency of versatile hand-held gadgets for remote interactive media spilling.

8.7 GREEN DATA CENTER TECHNOLOGIES

It is another innovation, and a vault for data stockpiling, data executives and data spread. Clients, frameworks, things make these data and so on. By managing various information and applications, data center (DC) overcomes large measures of vitality with high working expenses and huge CO_2 footprint. Besides, the age of enormous data is ascending by different universal things, for example, cell phones, sensors and so forth. In transit of the savvy world, the vitality productivity for DC turns out to be additionally squeezing [34,38]. Authors in [61] examined numerous procedures which upgraded the vitality utilization and forecast for DCs and their parts [61]. Notwithstanding the work of authors [61], authors in [64] introduced the improvement technique for the vitality effectiveness of DC with supporting quality of service (QoS). Besides, GDCs are giving data administrations to cloud-helped portable impromptu networks (MANET) in 5G [65]. Trendsetting innovations are used to limit structure paint and cover, supporting arrangement, alternative vitality. The examination in [66] gives a successful technique to lessen the force utilization without debasing the cooling effectiveness of DCs for greening IoT. The sparing vitality system in cloud data servers is diminishing, steering and looking through exchanges. People group et al. [67] investigated the systems coordinated viability into the vitality productive setting mindful specialist (e-CAB) structure to oversee cutting edge DCs. Nonetheless, the examination in [100] offers a GDC of cooling helped by cloud strategies, which comprise two subsystems: DC of the cooling framework, cloud the executives' stage. The DC of the cooling structure incorporates natural checking, cooling, correspondence, temperature control and ventilation, while the cloud stage gives information stockpiling, extensive data examination and forecast, and up-layer application. Insect province framework (ACS) based virtual machine (VM) is utilized for lessening the force utilization of DCs while protecting QoS necessities [68]. The value of utilizing ACS was to locate a close ideal arrangement. Moreover, dynamic VM is considered to lessen the vitality utilization of cloud DC while keeping up the ideal QoS [69]. Consequently, numerous clients share every gadget, and VM is utilized to use those physical gadgets. By reducing vitality, lack of VMs to QoS limits through data transmission, officials are talked into with subtleties in and for 5G networks. There are a few strategies used to upgrade vitality effectiveness for GDC, which can be accomplished from the accompanying viewpoints [34]: (a) utilize sustainable/green wellsprings of vitality; (b) utilize sufficient unique force the executives' technologies; (c) design more vitality proficient equipment procedures; (d) design epic vitality productive information center structures to accomplish energy protection; (e) construct skilled and precise data center force designs; (f) produce assistance from correspondence and registering strategies.

8.8 GREEN COMMUNICATION AND NETWORKING

Green wireless scenario assumes a pivotal job in green IoT. Green networking alludes to reasonable, energy mindful, productive and eco-friendly. Green network alludes to reduce CO_2 outflows, small presentation to radiation and vitality proficiency, regardless of the earlier proof given in [71], which designed a genetic calculation improvement for building the network up the network arranging. The idea supported by an examination that talks about how to increase the data rate is limiting CO2 discharges in intellectual WSNs.

Notwithstanding the effort of authors [74], Chan et al. [72] built up a lot of schemes for assessing the utilization of the energy consumption and CO_2 emanations of wireless environment administrations. The structuring of vehicular, specially appointed networks (VANETs) is designed to diminish vitality utilization [73]. The practicality of the mix of delicate and green is to explore through five associated zones of analysis (namely, vitality productivity and ghastly proficiency co-plan, re-examining flagging, no cells, imperceptible base stations and full-duplex radio) [75, 76]. Besides, Abrol et al. [77] introduced the impact and the developing technologies required for vitality effectiveness in the next generation networks (NGN). The requirement for receiving vitality productivity and CO_2 outflow is to satisfy the requests for expanding the limit, upgrading the rate of data and give immense QoS. Numerous sorts of analysts have been accomplished for sparing vitality by utilizing sun powered and improved QoS. Implementing the network coding-based technique and substantial stockpiling is valuable for saving power for smart IoT. Stochastic geometry technique for displaying different traffic examples can run effectively and accomplish a critical upgrade in vitality proficiency while keeping up QoS prerequisites [78]. Utility-based versatile obligation cycle (UADC) calculation has been framed to lessen delay, increment vitality proficiency and manage forever [79]. The hypertexts move convention is used to improve the lifetime and abbreviate the postponement for giving the unwavering quality [62]. These days, 5G may hope to affect our situation and life impressively as IoT vowed to form it proficient, agreeable. It depicts the significance of 5G innovation for improving the dependability and QoS between the machines, one another and humans. Moreover, 5G innovation empowers to give a considerable inclusion network, lessen the idleness, sparing vitality and bolster higher rate of data and framework limit. The 5G [105] utilization and its administrations for the public are indulging, in mechanical technology transmission, cooperation human and apply autonomy, coordination, eLearning, e-administration, open security, car, modern frameworks and so on.

8.9 GREEN MACHINE TO MACHINE TECHNOLOGY

As of late, machines are progressively getting more brilliant and ready to assemble information beyond human mediation. AI is the major technique in the improvement of numerous ongoing methodologies. Succeeding the possibility of the smart machine to machine (M2M) transmission [72] is essential to be utilized on an extensive scope. Tools ought to have a vast network to upgrade the cutting-edge PC machines and other electronic gadgets for accumulating the massive information. At that point, they can impart the ability to every single physical device and some other tools around. A machine speaks to an item that has electrical, mechanical, natural properties just as electronic properties. The advantage of such inherent radios' transmission is to ensure that M2M transmission is [87–89] sheltered and proficiently works for a wide range of assignments, for example, home, modern, clinical, just as business forms. The connection between machines is depicted [90]. In this manner, many devices can convey wisely, share data and work together dynamically [91]. M2M is the development adaptation of IoT, where machines speak with one another beyond human mediation. With the assistance of IoT, the billions of devices can interface, perceive, discuss and react to one another. Machine gadgets get to control (MDAC) methods accustomed for accomplishing the small vitality utilization (EC) and adjust to a variable conveyance of MDAC [92, 93].

8.10 GREEN CLOUD COMPUTING

Cloud computing (CC) is a developing virtualization innovation utilized over the internet. It gives ample of computational, boundless storage and administration conveyance employing the internet as adroitly. The joining of CC and IoT altogether has a full extent of research. The essential point of GCC is to advance the use of eco-accommodating items. Subsequently, Shuja et al. [94] introduced a thought of green registering with an emphasis on data technologies. Besides, Baccarelli et al. [95] tended to the green monetary arrangement in IoT over the current fog-bolstered network. Its major role is to diminish the utilization of risky things, expand vitality utilization and improve the recyclability.

Moreover, it could very well be accomplished by giving the side-effect life span asset distribution and paperless virtualization or appropriate force the executives. The thought is upheld by an investigation in [100], which talks about the different technologies for GCC by diminishing vitality utilization. Moreover, Sivakumar et al. [96] talked in depth about the combination of CC with IoT in various applications, designs, techniques, administration designs, database technologies, sensors and calculations. Additionally, Zhu et al. [97] formed a multi-strategy data conveyance (MMDD) for sensor-cloud (SC) clients, which accomplished less cost and conveyance duration. MMDD fuses four sorts of conveyance: transportation from WSN to SC clients, transport from cloud to SC clients, carriage from cloudlet to SC clients and transfer from SC clients to SC clients. The possibility of GCC is upheld by implementing the various methods to limit the forced necessity [98]. Authors in [98] found the significant specialized and investigated the force execution of GCC and GDC. Open and private mists were thought of and remembered vitality utilization for exchanging, data handling, transmission and data storage [99].

8.11 GREEN WIRELESS SENSOR NETWORK

A mix of wireless transmission and detecting has prompted the WSNs. A sensor is a mix of a colossal count of little, low-force and minimal effort electronic gadgets [86]. Countless sensors and base station (BS) hubs speak to the parts of WSNs.

Every hub comprises detecting, force and preparing and communication unit [86]. These hubs (sensors) are sent far and wide, estimating nearby and worldwide natural conditions, for example, climate, contamination and farming fields. Every sensor hub peruses from environmental factors, for example, temperature, sound, pressure, dampness, increasing speed and so forth. Sensors additionally speak with one another and convey the needful tactile data to BS utilizing impromptu innovation. They have restricted force and low preparing just as little storage limit, while a BS hub is legitimate. WSNs have various implications, for example, fire identification [107], object following [108], ecological monitoring and developing limitations in the military [109], control machine well-being monitoring and modern procedure monitoring [86].

Wake State Transmission Sensor Mode Reception Idle Sleep State Sensor modes for green IoT WSNs innovation needs to forward a sign productively and permit resting for negligible force use. Microchips in sensors should likewise have the option to wake and rest shrewdly. Thus, microchip patterns for WSNs incorporate decreasing vitality utilization, while speeding up. In this manner, green WSN is a developing idea where the life expectancy and throughput execution are amplified while the CO_2 outflow is sought after. The objective of WSN is providing adequate vitality to improve the framework forever and contribute to solid/strong transmission without trading off the general QoS. The possibility of vitality productive is bolstered by Mehmood et al. [101], who talked about the brilliant vitality productive of steering schemes communication for WSN concerning the plan exchange offs. Essentially, Rani et al. [106] contended with subtleties for progressive network plan and vitality proficient and adaptable IoT . Notwithstanding the work in [101], the authors in [102] featured the green WSN sensors to safeguard the life and directing over the WSN. Moreover, Yaacoub et al. [103] explored an agreeable methodology for sparing vitality for greening WSNs. Green WSN for empowering green IoT was bolstered by an investigation in [104], which concentrated on expanding vitality productivity, broadening network lifetime, lessening hand-off hubs and decrease in framework financial plan. The work was actualized in four stages which were: the production of various leveled framework structures and arrangement of sensor/actuator hubs, grouping the hubs, formation of improvement model to acknowledge green IoT and finally the count of insignificant vitality among the hubs. The discoveries demonstrated that the designed scheme was flexible, vitality sparing and cost effective when compared with the current WSNs arrangement plans. In this way, it is appropriate for green IoT.

8.12 GREEN RFID TECHNOLOGY

RFID is the joined terminology of RF and ID where RF alludes to wireless communication innovation, and ID implies label ID data. It is assumed as one of the prominent wireless transmission frameworks to empower the IoT. Moreover, it need not bother with view (LoS) and can delineate genuine world into the visual world simple [80]. What is more, RFID is an automated information assortment, and empowering articles to interface through networks those utilize radio waves to recover, recognize and accumulate information remotely. The utilization of electromagnetic in the radio recurrence and the utilization of savvy standardized identifications to follow things in a store are alluding to RFID fuse. The characterization of RFID is inactive and dynamic . Inactive things have not batteries on the board, and the communication recurrence is limited. Then again, dynamic RFID incorporates batteries that give power to the communicating signal. RFID assumes an indispensable job that causes the world to be greener by lessening the emanations of the vehicle, sparing vitality utilized and cultivate waste removal and so forth. Amin proposed an answer for the green RFID radio wires for inserted sensors [81]. The plan adaptability reception apparatus ingeniously gives the satisfactory alignment of the dampness sensor following characterized prerequisites [82]. The segments of RFID are tag (conveying distinguishing proof data), per user (perusing/composing label data and interfaces with a framework), receiving wire and station (process the data) [83]. Hubbard et al. [84] concentrated on upgrading the lifelong of unmanned elevated vehicle (UAV) battery and RFID per user identification extend. The mixes of UAV and RFID are to give extra data that can be actualized in flexibly chain the board frameworks. Energize a multipurpose RFID label utilizing UAV in a natural monitoring activity was talked about in [85]. Moreover, Choi et al. [34] considered a UAV indoor limitation method utilizing uninvolved UHF far-field RFID frameworks. UAV restriction and following have been thought about for accomplishing effortlessness and cost productivity. UAV is utilized for data assortment from RFID sensors by means of downloading projected data, legitimately moving towards them and hovering over them [35].

8.13 MACHINE LEARNING-BASED GREEN ICT

Data science is the blend of various streams of sciences that utilize data mining, ML and different methods to discover designs and experiences from information. These methods incorporate an expansive scope of algorithms in various areas. The way toward applying data investigation strategies to specific territories includes characterizing data types, for example, volume, assortment, speed, data models, for example, neural systems, order, bunching techniques and applying proficient algorithms that coordinate with data attributes.

ML has its inceptions in computer science; there have been a few vector quantization techniques [111] created in broadcast communications and sign preparing for coding and pressure [110]. In computer and data science, learning is practiced depending on models (data tests) and experience.

Ordinarily, the feature extraction phase will remove smaller data-bearing boundaries that can portray the information. The grouping stage should be prepared by a ML calculation to perceive and characterize the assortment of highlights. The field of ML is tremendous, and applications are overgrowing, particularly with the development of quick cell phones that likewise approach distributed computing [112]. Packing and removing data from sensors and big data have, as of late, raised enthusiasm for the zone. Shrewd city ventures, portable well-being observing, organized security, producing, self-propelled autos, observation, insightful fringe control; each application has its mannerisms and needs highlights, versatile study and data combination. Data pressure, factual sign and information interpretations have a tremendous job transmitting and deciphering data and delivering necessary examination. The classification of ML algorithms depends on the attributes, method of learning and how information is utilized: directed, unaided and semi-managed schemes. This kind of characterization is significant in recognizing the job of the information, schemes utilization and learning designs comparative with the implications.

IoT [113] is an arrangement of associated physical gadgets, keen machines or items that have remarkable identifiers. Equipment comprises hardware, programming, sensors and radios, empowering these articles to consistently gather and move data. Sensors that comprise a transducer will change over some type of physical procedure into an electrical sign. Models incorporate mouthpieces, cameras, accelerometers, thermometers, pressure sensors and so on. Maybe a cell phone is a genuine case of an associated gadget that installs a few heterogeneous sensors, including receiver clusters, at any rate, two cameras, magnetometers, accelerometers and so on. Original PDAs, for instance, usually included six sensors. Accelerometers and magnetometers have been utilized in numerous applications, including machine checking, basic observing, human action and social insurance [114]. Different zones of collective detecting and ML incorporate restriction [117].

Sharp amusement and data trade frameworks, for example, intelligent speakers, join different advancements, for example, roundabout amplifier exhibits, nearby and cloud-based machine study and data recovery strategies. The Amazon Echo speaks to an ongoing case of an IoT gadget that has a roundabout receiver cluster alongside voice acknowledgment abilities. Nearby and distributed computing permit this gadget to interface with different frameworks, trade data, give e-administrations, playback music and news on request, and provide human to a machine interface to a brilliant home [118]. The interconnection of IoT shrewd gadgets is additionally empowering propelled enormous scope applications, for example, brilliant urban communities [115], huge scope savvy systems and radios, keen grounds frameworks [116]. The field of sensors and IoT applications is huge, and enormous scope applications are starting to rise. These incorporate a few savvy and associated well-being and network frameworks.

8.14 CHALLENGES IN MACHINE LEARNING-BASED GREEN ICT

- Lack of perception about green ICT is perhaps the greatest obstruction in achieving the green IT.
- Present days have adopted diverse kind of paradigms and technologies to assemble the computers. So, it is one of the difficulties, without losing the effectiveness and determinations of PC in what capacity can be actualizing green IT.
- Many organizations are putting away those PCs that are no more much operational. Since they are so obsolete PCs, programming to reuse them is a major challenge.
- Personal data to fall into an inappropriate hand; however, attorneys should likewise assume the protection of their customer's data. As certain cases emerged, in 2007 a digital security master got a $500 PC at a trade amass. By utilizing software, he had the option to extract the information from the PC, which had been utilized by a home loan organization, and gained the credit reports of about 300 individuals.

Malicious Use of Machine Learning 131

- Designing a computer to be more environment-friendly and productive at the same time is a great challenge [121]. As designing involves the innovation in productivity, speed, adaptability, perseverance and different specifications of PC.
- Control on expanding necessities of heat evacuating device, which rises because of rise in absolute energy utilization by IT device and the ML schemes with high dataset.

8.15 MALICIOUS USE OF MACHINE LEARNING IN GREEN ICT

No organization or client will ever wish their information or any close data to fall into an inappropriate hand. As certain cases were emerged, in 2018 a digital security master got a $700 PC at a trade gather. By utilizing simple software/hacking, he had the option to extract the information and details of the company from the PC, which had been utilized by a home loan organization. Too, he was additionally able to get the access of client names and passwords the organization's representatives had used to get to credit authorities. Another malicious use occurred latently when an organization bought 25 PCs from secondhand shops from different communities. The TV program team additionally obtained four PCs for nothing at a town dump. The PCs were given to a specialist who found that none of the previous users of the PCs had cleaned the data. The machine incorporates money related and individual data that could have been utilized to submit wholesale fraud and misrepresentation. ML techniques work adroitly, and intruder can use any of the techniques to hack the system. So, security issue is an enormous hindrance in the execution of green IT with ML. Since data can be restored and used by repairing the issue of hard drive ROM or RAM. Notwithstanding the data center cooling [123] down is a significant issue. Data centers are essential spine of any figuring association and must be precise, dependable and accessible inevitably.

8.16 SOLUTION FOR MALICIOUS USE OF MACHINE LEARNING IN GREEN ICT

- Build a Green Data Center for Reducing Energy
  Data center stakes so much energy [63] to operate itself, and in previous years the cost of the energy usage by data centers are too high to bear. But this cannot be avoided as the reliability, capacity supply and others issues of the data centers have been minimized with the increment of the cost. So, data centers require their own specific strategy for being green. Here is a little approach to make the data centers green. This is depending on the daily usage bill as well as the monthly usage bill.
- Efficient Servers Usage by Virtualization:
  Virtual machines [124] use is increasing day by day as people are using many smart gadgets which are highly computing and require virtual machines for their smooth operation. Virtualization is the basic technique of creating various resources from the available physical infrastructure. It is the backbone of cloud computing technology. It is a technique which directly impacts

the environment. Virtualization was introduced by IBM in 1960 but it was implemented by Windows x86 in 1990s. By using virtualization, a system administrator could combine several physical systems into virtual machines on one single, powerful system, thereby unplugging the original hardware and reducing power and cooling consumption. For implementation of such kind of virtualization, Intel Corporation and AMD have built proprietary virtualization enhancements.

- Information Resource Tier Optimization:
  The information resource tier represents vital database management systems in the universal computation world. General paradigms include databases, directories, filesystems and flat files. ML algorithms run on a large dataset which not only provide the optimized solution but also provide robust information. So smart and highly encrypted servers should be set up which should save energy consumption by using low power [122] batteries (spongy silicon-based electrode Li-ion, sodium-ion and silver-zinc, etc.). Simultaneously, data stored on the servers should be highly confidential and encrypted.
- Improved Data Center Cooling Methods:
  This is achieved by enhancing the data center cooling configuration of machines and minimizing considerable amount of energy leaks. It can make efficient data centers by following new practices in data center layouts and racks.
- Green Software:
  Recently, green software inclination has become a research subject for most of the software developer companies because of the need for sustainable development. Most of the research has been done on the characterization, metrics and technical answer for green software, but few have addressed green software from the business and company perspective. Business organizations are moving toward green software and still some important steps need to be taken.

8.17 APPLICATIONS OF GREEN ICT

Critical changes in our surroundings have happened, and a few upgradations will happen soon on account of the advancements in IoT. Be that as it may, the expense of the progress is conceivably critical because of the expansion in e-squander, risky discharges and vitality use. Green IoT is assessed to roll out significant improvements soon and would prompt a green domain. Green IoT functions are centered on sparing vitality, lessening CO_2 release and contaminated and dangerous [34].

- Smart home: A green IoT empowers home-prepared warming, illumination and automatic gadgets which are guarded remotely by a PC/cell phone. The focal portable/PC in-house acknowledges voice orders. It recognizes inhabitants for customized activities and reactions, television, PC and telephone converge into one gadget and so forth. Consideration should be given to the current pattern of green IoT which includes green structure, green use, green creation, lastly green removal/reusing; to diminish the effect on the earth [35].
- Industrial automation: These have been computerized with machines that can accomplish the task completely and accurately. The mechanical robotization

dependent on green IoT is portrayed quickly in [36]. Smart medical services allude to the execution of various activities in patients for catching, observing and following the human body [37]. Presenting innovative and propelled sensors associated with the web for creating necessary information is the IoT upheaval in the human services industry [86]. The subsequent accomplishments of proficient social services administrations are upgrading the consideration quality, improving access to the mind and diminishing consideration costs.

- Smart Grid: The utilization of correspondence sensor based IoT, and savvy framework is talked about in [39]. Yang et al. [40] proposed an ease remote memory validation for the shrewd matrix. Besides, Liu et al. [41] suggested a few ways to deal with increment information legitimacy of IoT level information misfortune for savvy urban communities. The future brilliant network can gauge and offer vitality utilization and fabricate full vitality frameworks.
- Smart urban communities: IoT can be described by effective vitality use to empower an intelligent, practical world [42, 43]. Thus, the machines are proposed to be outfitted with extra tactile and correspondence additional items to make the world more astute. Devices can detect objects in a city and talk to each other. The smart city incorporates keen leaving [44], savvy light [45], high air, [46], bright traffic the executives [47]. Maksimovic et al [48] summed up the way to novel innovation and big information achievement in enthusiastic urban communities, where near low contamination will improve personal satisfaction.. Savvy and associated networks have developed from the idea of interested in urban communities [49]. The standards and uses of shrewd urban areas are talked about in [50].
- Smart agriculture: It will help the ranchers to battle with the immense difficulties which they suffer. Nandyala et al. [51] introduced the use of the green IoT considering horticulture. Green IoT and nanotechnology show up as acceptable answers to make the maintainable agribusiness and food industry [52].

8.18 CONCLUSION

In this chapter, an attempt is made to analyze the ML, its techniques to figure out the classification, regression and clustering problems. Different kinds of machine learning and green ICT techniques have been discussed. Notwithstanding, the ML is incorporated with green ICT examining its advantages, challenges and solutions. It is expected that this chapter will illuminate the readers to take an informed decision in distinguishing the accessible alternatives of ML algorithms with green ICT.

REFERENCES

[1] S. Ray, A quick review of machine learning algorithms, 2019 International Conference on Machine Learning, Big Data, Cloud and Parallel Computing (COMITCon), Faridabad, India, 2019, pp. 35–39.

[2] R. Nagar, Y. Singh, A literature survey on machine learning algorithms, International Journal of Emerging Technologies and Innovative Research, 6 (4) (April) (2019): 471–474.

[3] M. Awad, R. Khanna, *Efficient Learning Machines: Theories, Concepts, and Applications for Engineers and System Designers*, Springer, 2015, pp. 268.
[4] L. von Rueden, S. Mayer, K. Beckh, B. Georgiev, S. Giesselbach, R. Heese, B. Kirsch et al. Informed machine learning—A taxonomy and survey of integrating knowledge into learning systems." arXiv preprint arXiv:1903.12394 (2019).
[5] J. Watt, R. Borhani, A. Katsaggelos, *Machine Learning Refined: Foundations, Algorithms, and Applications*, Cambridge University Press, 2016.
[6] K. Das, R. Narayan Behera, A survey on machine learning: concept, algorithms and applications, *International Journal of Innovative Research in Computer and Communication Engineering* 5 (2017), pp. 1301–1309.
[7] P. Domingos. A few useful things to know about machine learning. *Communications of the ACM* 55(10) (2012), 78–87.
[8] Y. Singh, P.K. Bhatia, O. Sangwan, A review of studies in machine learning technique, *International Journal of Computer Science and Security* 1 (2007): 70–84.
[9] K. Solanki and A. Dhankar. A review on Machine Learning Techniques. International Journal of Advanced Research in Computer Science 8(3) (2017).
[10] I. Spectrum, Will the future of AI learning depend more on nature or nurture? https://spectrum.ieee.org/tech-talk/robotics/artificial-intelligence/ai-and-psychology-researchersdebate-the-future-of-deep-learning.
[11] M.H. Fazel Zarandi, E. Hadavandi, B. Turksen, A hybrid fuzzy intelligent agent-based system for stock price prediction, *International Journal of Intelligent Systems* 11(2012): 1–23.
[12] V. Sharma, S. Rai, A. Dev, A comprehensive study of artificial neural networks, *International Journal of Advanced Research in Computer Science and Software Engineering*, ISSN 2277128X, 2 (10) (October) (2012): 5.
[13] T. Xiuyi1, G. Yuxia1, Research on application of machine learning in data mining. In IOP conference series: materials science and engineering, vol. 392, no. 6, p. 062202, IOP Publishing, 2018.
[14] F.E. Bock, R.C. Aydin, C.J. Cyron, N. Huber, S.R. Kalidindi, B. Klusemann. A review of the application of machine learning and data mining approaches in continuum materials mechanics. *Frontiers in Materials* 6 (2019), p. 110.
[15] S. Murugesan, Harnessing green IT: principles and practices, *IT Professional* 10 (1) (2008): 7.
[16] L. Lakhani, Green computing—a new trend in IT, *International Journal of Scientific Research in Computer Science and Engineering* 4 (3) (2016): 11–13.
[17] K. Nanath, R.R. Pillai. The influence of green IS practices on competitive advantage: mediation role of green innovation performance, *Information Systems Management* 34 (1) (2017): 3–19.
[18] J. Brownlee, Probability for machine learning: Discover how to harness uncertainty with Python. Machine Learning Mastery, 2019.
[19] J. Alzubi, A. Nayyar, A. Kumar. Machine learning from theory to algorithms: an overview. In Journal of physics: conference series, vol. 1142, No. 1, IOP Publishing, 2018, pp. 012012.
[20] M. Praveena, V. Jaiganesh, Literature review on supervised machine learning algorithms and boosting process, *International Journal of Computer Applications*, ISSN No. 0975-8887, 169 (2017): 8.
[21] E. Angelino, M.J. Johnson, R.P. Adams. Patterns of scalable Bayesian inference, *Foundations and Trends in Machine Learning* 9 (2–3) (2016): 119–247.
[22] K. Chen, C. Wang, Support vector regression with genetic algorithms in forecasting tourism demand, *Tourism Management* 28 (2007): 215–226.

[23] Y. Wei-Chiang Hong, L. Dong, S, Chen, SVR with hybrid chaotic genetic algorithms for tourism demand forecasting, *Applied Soft Computing* 11 (2011): 1881–1890.
[24] R.T. Hastie, J. Friedman, Unsupervised learning, *The Elements of Statistical Learning*, Springer, New York, 2009, 485–585.
[25] O. Simeone, A very brief introduction to machine learning with applications to communication systems. *IEEE Transactions on Cognitive Communications and Networking* 4(4) (2018): 648–664.
[26] X. Zhu, A.B. Goldberg, Introduction to semi–supervised learning, Synthesis Lectures on Artificial Intelligence and Machine Learning 3 (1) (2009): 1–130.
[27] X. Zhu, Semi-supervised learning literature survey, Computer Sciences, University of Wisconsin-Madison, Technical Report No. 1530, (2005).
[28] R.S. Sutton, A.G. Barto, Reinforcement Learning: An Introduction, MIT Press, 2018.
[29] S.B. Hiregoudar, K. Manjunath, K.S. Patil, A survey: research summary on neural networks, *International Journal of Research in Engineering and Technology*, ISSN: 2319 1163, 03 (Special issue) (03) (May) (2014): 385–389.
[30] https://www.edureka.co/executive-programs/machine-learning-and-ai?utm_source=search&utm_medium=google&utm_campaign=PGDINML-Search-ML&utm_term=ML-Course&utm_content=rsa&gclid=EAIaIQobChMIrKuiz9258gIVk4ZLBR3NhwHEEAAYASAAEgLu3_D_BwE
[31] https://en.wikipedia.org/wiki/Instance-based_learning.
[32] T. Liu, Z. Chen, C. Chen, Z. Duan, B.A. Zhao, K. Dynamic, Nearest neighbor map matching method combined with neural network, 2019 IEEE Intelligent Transportation Systems Conference (ITSC), Oct 27, 2019, pp. 3573–3578.
[33] P. Harrington, *Machine Learning in Action*, Manning Publications Co., Shelter Island, New York, ISBN 9781617290183, 2012.
[34] C. Zhu, V.C. Leung, L. Shu, E.C.-H. Ngai, Green internet of things for the smart world, *IEEE Access*, 3 (2015): 2151–2162.
[35] S. Aslam, N.U. Hasan, A. Shahid, J.W. Jang, K.-G. Lee, Device centric throughput and QoS optimization for IoTs in a smart building using CRN-techniques, *Sensors* 16 (2016): 1647.
[36] N. Kulkarni, S. Abhang, Green industrial automation based on IOT: a survey, *International Journal of Emerging Trends in Science and Technology* 4 (2017): 5805–5810|.
[37] Z.M. Kalarthi, A review paper on smart health care system using internet of things, *International Journal of Research in Engineering and Technology (IJRET)* 5 (2016): 80–83.
[38] Y. Liu, X. Weng, J. Wan, X. Yue, H. Song, A.V. Vasilakos, Exploring data validity in transportation systems for smart cities, *IEEE Communications Magazine* 55 (2017): 26–33.
[39] H. Yanti, The applications of WiFi-based wireless sensor network in internet of things and smart grid, *Bulletin Innovis ICT & Ilmu Komputer*, 2 (2015).
[40] X. Yang, X. He, W. Yu, J. Lin, R. Li, Q. Yang, et al., Towards a low-cost remote memory attestation for the smart grid, *Sensors* 15 (2015): 20799–20824.
[41] S. Bhattacharjee, P. Roy, S. Ghosh, S. Misra, M.S. Obaidat, Wireless sensor network-based fire detection, alarming, monitoring and prevention system for Bord-and-Pillar coal mines, *Journal of Systems and Software* 85 (2012): 571–581.
[42] R. Petrolo, V. Loscrì, N. Mitton, Towards a smart city based on cloud of things: a survey on the smart city vision and paradigms, Transactions on Emerging Telecommunications Technologies 28(1), (2015), p. e2931.

[43] P. Sathyamoorthy, E.C.-H. Ngai, X. Hu, V.C. Leung, Energy efficiency as an orchestration service for mobile internet of things, cloud computing technology and science (CloudCom), 2015 IEEE 7th International Conference on IEEE (2015): 155–162.

[44] P. Ramaswamy, IoT smart parking system for reducing greenhouse gas emission, 2016 IEEE International Conference on Recent Trends in Information Technology (ICRTIT), 2016, pp. 1–6.

[45] M. Popa, A. Marcu, A solution for street lighting in smart cities, Carpathian Journal of Electronic and Computer Engineering 5 (2012): 91.

[46] A.M. Vegni, M. Biagi, R. Cusani, Smart vehicles, technologies and main applications in vehicular ad hoc networks, In: L. Galati Giordano and L. Reggiani (eds.), *Vehicular Technologies-Deployment and Applications*, InTech, Croatia, (2013). pp. 3–20.

[47] K. Su, J. Li, H. Fu, Smart city and the applications, 2011 IEEE International Conference on Electronics, Communications and Control (ICECC), 2011, pp. 1028–1031.

[48] M. Maksimovic, The role of green internet of things (G-IoT) and big data in making cities smarter, safer and more sustainable, *International Journal of Computing and Digital Systems* 6 (2017): 175–184.

[49] Y. Sun, H. Song, A.J. Jara, R. Bie, Internet of things and big data analytics for smart and connected communities, *IEEE Access*, 4 (2016): 766–773.

[50] H. Song, R. Srinivasan, S. Jeschke, T. Sookoor, Smart cities: foundations, principles and applications, John Wiley & Sons, 2017, *IEEE Internet of Things Journal*, 3 (2016): 1437–1447.

[51] C.S. Nandyala, H.-K. Kim, Green IoT agriculture and healthcare application (GAHA), *International Journal of Smart Home* 10 (2016): 289–300.

[52] M. Maksiimovici, E. Omanovic-Miklicanin, Green internet of things and green nanotechnology role in realizing smart and sustainable agriculture VIII International Scientific Agriculture Symposium "AGROSYM 2017", Jahorina, Bosnia and Herzegovina, 2017.

[53] F.K. Shaikh, S. Zeadally, Energy harvesting in wireless sensor networks: a comprehensive review, *Renewable and Sustainable Energy Reviews* 55 (2016): 1041–1054.

[54] A. Adelin, P. Owezarski, T. Gayraud, On the impact of monitoring router energy consumption for greening the internet, 2010 11th IEEE/ACM International Conference on Grid Computing (GRID), IEEE, 2010, pp. 298–304.

[55] M. Baldi, Y. Ofek, Time for a "greener" internet, 2009 IEEE International Conference on Communications Workshops, 2009. ICC Workshops, IEEE, 2009, pp. 1–6.

[56] Y. Suh, K. Kim, A. Kim, Y. Shin, A study on impact of wired access networks for green internet, *Journal of Network and Computer Applications* 57 (2015): 156–168.

[57] Y. Suh, J. Choi, C. Seo, Y. Shin, A study on energy savings potential of data network equipment for a green internet, 2014 16th International Conference on Advanced Communication Technology (ICACT) IEEE, 2014, pp. 1146–1151.

[58] Y. Yang, D. Wang, M. Xu, S. Li, Hop-by-hop computing for green internet routing, 2013 21st IEEE International Conference on Network Protocols (ICNP) IEEE, 2013, pp. 1–10.

[59] Y. Yang, D. Wang, D. Pan, M. Xu, Wind blows, traffic flows: green internet routing under renewable energy, 2016-The 35th Annual IEEE International Conference on Computer Communications, IEEE INFOCOM, IEEE, 2016, pp. 1–9.

[60] M.A. Hoque, M. Siekkinen, J.K. Nurminen, Energy efficient multimedia streaming to mobile devices—a survey, *IEEE Communications Surveys & Tutorials*, 16 (2014): 579–597.

[61] M. Dayarathna, Y. Wen, R. Fan, Data center energy consumption modeling: a survey, *IEEE Communications Surveys & Tutorials*, 18 (2016): 732–794.

[62] T. Himsoon, W.P. Siriwongpairat, Z. Han, K.R. Liu, Lifetime maximization via cooperative nodes and relay deployment in wireless networks, *IEEE Journal on Selected Areas in Communications*, 25 (2007): 306–317.

[63] M. Zhou, Q. Cui, R. Jantti, X. Tao, Energy-efficient relay selection and power allocation for two-way relay channel with analog network coding, *IEEE Communications Letters*, 16 (2012): 816–819.

[64] N. Cordeschi, M. Shojafar, D. Amendola, E. Baccarelli, Energy-efficient adaptive networked datacenters for the QoS support of real-time applications, *The Journal of Supercomputing*, 71 (2015): 448–478.

[65] N.D. Han, Y. Chung, M. Jo, Green data centers for cloud-assisted mobile ad hoc networks in 5G, *IEEE Network*, 29 (2015): 70–76.

[66] A. Roy, A. Datta, J. Siddiquee, B. Poddar, B. Biswas, S. Saha, et al., Energy-efficient data centers and smart temperature control system with IoT sensing, 2016 IEEE 7th Annual Information Technology, Electronics and Mobile Communication Conference (IEMCON), IEEE, 2016, pp. 1–4.

[67] C. Peoples, G. Parr, S. McClean, B. Scotney, P. Morrow, Performance evaluation of green data centre management supporting sustainable growth of the internet of things, *Simulation Modelling Practice and Theory*, 34 (2013): 221–242.

[68] F. Farahnakian, A. Ashraf, T. Pahikkala, P. Liljeberg, J. Plosila, I. Porres, et al., Using ant colony system to consolidate VMs for green cloud computing, *IEEE Transactions on Services Computing*, 8 (2015): 187–198.

[69] P. Matre, S. Silakari, U. Chourasia, Ant colony optimization (ACO) based dynamic VM consolidation for energy efficient cloud computing, *International Journal of Computer Science and Information Security*, 14 (2016): 345.

[70] E. Baccarelli, D. Amendola, N. Cordeschi, Minimum-energy bandwidth management for QoS live migration of virtual machines, *Computer Networks*, 93 (2015): 1–22.

[71] G. Koutitas, Green network planning of single frequency networks, *IEEE Transactions on Broadcasting*, 56 (2010): 541–550.

[72] C.A. Chan, A.F. Gygax, E. Wong, C.A. Leckie, A. Nirmalathas, D.C. Kilper, Methodologies for assessing the use-phase power consumption and greenhouse gas emissions of telecommunications network services, *Environmental Science & Technology* 47 (2012): 485–492.

[73] W. Feng, H. Alshaer, J.M. Elmirghani, Green information and communication technology: energy efficiency in a motorway model, *IET Communications*, 4 (2010): 850–860.

[74] Z. Alavikia, A. Ghasemi, Collision-aware resource access scheme for LTE-based machine-to-machine communications, *IEEE Transactions on Vehicular Technology* 67(5) (2018): 4683–4688.

[75] R. Lu, X. Li, X. Liang, X. Shen, X. Lin, GRS: The green, reliability, and security of emerging machine to machine communications, *IEEE Communications Magazine*, 49 (2011): 28–35.

[76] G. Mao, 5G green mobile communication networks, *China Communications*, 14 (2017): 183–184.

[77] A. Abrol, R.K. Jha, Power optimization in 5G networks: a step towards GrEEn communication, *IEEE Access*, 4 (2016): 1355–1374.

[78] L. Zhou, Z. Sheng, L. Wei, X. Hu, H. Zhao, J. Wei, et al., Green cell planning and deployment for small cell networks in smart cities, *Ad Hoc Networks*, 43 (2016): 30–42.

[79] J. Wang, C. Hu, A. Liu, Comprehensive optimization of energy consumption and delay performance for green communication in internet of things, *Mobile Information Systems* 2017 (2017).

[80] F.K. Shaikh, S. Zeadally, E. Exposito, Enabling technologies for green internet of things, *IEEE Systems Journal*, 11(2) (2015), 983–994.
[81] C. Xiaojun, L. Xianpeng, X. Peng, IOT-based air pollution monitoring and forecasting system, 2015 International Conference on Computer and Computational Sciences (ICCCS), IEEE, 2015, pp. 257–260.
[82] Y. Amin, Printable green RFID antennas for embedded sensors, KTH Royal Institute of Technology (2013). Retrieved from diva-portal.org
[83] Y. Amin, R.K. Kanth, P. Liljeberg, A. Akram, Q. Chen, L.-R. Zheng, et al., Printable RFID antenna with embedded sensor and calibration functions, Proceedings of the Progress in Electromagnetics Research Symposium, Stockholm, Sweden, 2013, pp. 567–570.
[84] P.J. Zelbst, V.E. Sower, K.W. Green Jr, R.D. Abshire, Radio frequency identification technology utilization and organizational agility, *Journal of Computer Information Systems*, 52 (2011): 24–33.
[85] B. Hubbard, H. Wang, M. Leasure, Feasibility study of UAV use for RFID material tracking on construction sites, Proceedings of the 51st ASC Annual International Conference Proceedings, College Station, TX, USA, 2016.
[86] M. Allegretti, S. Bertoldo, Recharging RFID tags for environmental monitoring using UAVs: a feasibility analysis, *Wireless Sensor Network*, 7 (2015): 13.
[87] B. Prabhu, N. Balakumar, A. Antony. Wireless sensor network based smart environment applications. *Wireless Sensor Network Based Smart Environment Applications* (January 31). IJIRT 3, no. 8 (2017).
[88] D.P. Van, B.P. Rimal, S. Andreev, T. Tirronen, M. Maier, Machine-to-machine communications over FiWi enhanced LTE networks: a power-saving framework and end-to-end performance, *Journal of Lightwave Technology* 34 (2016): 1062–1071.
[89] L. Sun, H. Tian, L. Xu, A joint energy-saving mechanism for M2M communications in LTE-based system, 2013 IEEE Wireless Communications and Networking Conference (WCNC), IEEE, 2013, pp. 4706–4711.
[90] S. Ajah, A. Al–Sherbaz, S. Turner, P. Picton, Machine–to–machine communications energy efficiencies: the implications of different M2M communications specifications, *International Journal of Wireless and Mobile Computing* 8 (2015): 15–26.
[91] R.S. Cheng, C.M. Huang, G.S. Cheng, A congestion reduction mechanism using D2D cooperative relay for M2M communication in the LTE-a cellular network, *Wireless Communications and Mobile Computing* 16 (2016): 2477–2494.
[92] L. Ji, B. Han, M. Liu, H.D. Schotten, Applying device-to-device communication to enhance IoT services, arXiv preprint arXiv:1705.03734, 1(2) (2017), 85–91.
[93] F. Hussain, A. Anpalagan, M. Naeem, Multi-objective MTC device controller resource optimization in M2M Communication, 2014 27th Biennial Symposium on Communications (QBSC), IEEE, 2014, pp. 184–188.
[94] Y. Liu, Z. Yang, R. Yu, Y. Xiang, S. Xie, An efficient MAC protocol with adaptive energy harvesting for machine-to-machine networks, *IEEE Access*, 3 (2015): 358–367.
[95] Z. Abbas, W. Yoon, A survey on energy conserving mechanisms for the internet of things: wireless networking aspects, *Sensors*, 15 (2015): 24818–24847.
[96] E. Baccarelli, P.G.V. Naranjo, M. Scarpiniti, M. Shojafar, J.H. Abawajy, Fog of everything: energy-efficient networked computing architectures, research challenges and a case study, *IEEE Access*, 5 (2017): 9882–9910.
[97] S. Sivakumar, V. Anuratha, S. Gunasekaran, Survey on integration of cloud computing and internet of things using application perspective, *International Journal of Emerging Research in Management &Technology* 6 (2017): 101–108.
[98] C. Zhu, V.C. Leung, K. Wang, L.T. Yang, Y. Zhang, Multi-method data delivery for green sensor-cloud, *IEEE Communications Magazine* 55 (2017): 176–182.

[99] A. Jain, M. Mishra, S.K. Peddoju, N. Jain, Energy efficient computing-green cloud computing, 2013 International Conference on Energy Efficient Technologies for Sustainability (ICEETS), IEEE, 2013, pp. 978–982.
[100] J. Baliga, R.W. Ayre, K. Hinton, R.S. Tucker, Green cloud computing: balancing energy in processing, storage, and transport, *Proceedings of the IEEE*, 99 (2011): 149–167.
[101] J. Azevedo, F. Santos, Energy harvesting from wind and water for autonomous wireless sensor nodes, *IET Circuits, Devices & Systems* 6 (2012): 413–420.
[102] A. Mehmood, H. Song, Smart energy efficient hierarchical data gathering protocols for wireless sensor networks, *SmartCR* 5 (2015): 425–462.
[103] P.G.V. Naranjo, M. Shojafar, H. Mostafaei, Z. Pooranian, E. Baccarelli, P-SEP: a prolong stable election routing algorithm for energy-limited heterogeneous fog-supported wireless sensor networks, *The Journal of Supercomputing*, 73 (2017): 733–755.
[104] E. Yaacoub, A. Kadri, A. Abu-Dayya, Cooperative wireless sensor networks for green internet of things, Proceedings of the 8th ACM Symposium on QoS and Security for Wireless and Mobile Networks, ACM, 2012, pp. 79–80.
[105] R.V. Rekha, J.R. Sekar, A unified deployment framework for realization of green internet of things (GIoT), *Middle East Journal of Scientific Research*, 24 (2016): 187–196.
[106] D. Amendola, N. Cordeschi, E. Baccarelli, Bandwidth management VMs live migration in wireless fog computing for 5G networks, 2016 5th IEEE International Conference on Cloud Networking (Cloudnet), IEEE, 2016, pp. 21–26.
[107] S. Rani, R. Talwar, J. Malhotra, S.H. Ahmed, M. Sarkar, H. Song, A novel scheme for an energy efficient internet of things based on wireless sensor networks, *Sensors*, 15 (2015): 28603–28626.
[108] J. Lloret, M. Garcia, D. Bri, S. Sendra, A wireless sensor network deployment for rural and forest fire detection and verification, *Sensors* 9 (2009): 8722–8747.
[109] F. Viani, L. Lizzi, P. Rocca, M. Benedetti, M. Donelli, A. Massa, Object tracking through RSSI measurements in wireless sensor networks, *Electronics Letters* 44 (2008): 653–654.
[110] S.R.B. Prabhu, N. Balakumar, A. Johnson Antony. Evolving constraints in military applications using wireless sensor networks, *International Journal of Innovative Research in Computer Science & Technology (IJIRCST)* (2017): 2347–5552.
[111] L. Yoseph, B. Andrés Gray, Robert M. An algorithm for vector quantization, *IEEE COM* 28 (1) (Jan.) (1980): 84–95.
[112] A.S. Spanias, Speech coding: a tutorial review, *Proceedings of the IEEE*, 82 (10) (October) (1994): 1441–1582.
[113] E.G. Ularu et al., Mobile computing and cloud maturity—introducing machine learning for ERP configuration automation, *Informatica Economica*, 17 (1/2013) (Mar.) (2013): 40–52.
[114] Gubbi, J. et al., Internet of things (IoT): a vision, architectural elements, and future directions, *Future Generation Computer Systems* 29 (7) (2013): 1645–1660.
[115] C. Aldrich, L. Auret, *Unsupervised Process Monitoring and Fault Diagnosis with Machine Learning Methods*, vol. 16, no. 3. London: Springer, 2013.
[116] Hwang, J.-S., Choe, Y.H., Smart cities Seoul: a case study (PDF). ITU-T Technology Watch (February 2013). Retrieved on 23 October 2016.
[117] Sensor networks and the smart campus, 2014. [Online]. Available: https://beaverworks.ll.mit.edu/CMS/bw/smartcampusfuture. Accessed: Dec. 12, 2016.
[118] Miller, S., Zhang, X., Spanias, A. Multipath effects in GPS receivers. *Synthesis Lectures on Communications*, 8 (1) (2015): 1–70.
[119] www.ifixit.com/Teardown/Amazon+Echo+Teardown.
[120] www.Batteryuniversity.com/ultra_fast_chargers.

[121] N.S. More, R.B. Ingle, Challenges in green computing for energy saving techniques, 2017 International Conference on Emerging Trends & Innovation in ICT (ICEI), Pune, 2017, pp. 73–76, doi:10.1109/ETIICT.2017.7977013.

[122] Á.M. Groba, P.J. Lobo, M. Chavarrías, Closed-loop system to guarantee battery lifetime for mobile video applications, *IEEE Transactions on Consumer Electronics*, 65 (1) (2019): 18–27.

[123] D. Gmach, Y. Chen, A. Shah, J. Rolia, C. Bash, T. Christian, et al., Profiling sustainability of data centers, 2010 IEEE International Symposium on Sustainable Systems and Technology (ISSST), 2010, pp. 1–6.

[124] A. Rahman, X. Liu, F. Kong, A survey on geographic load balancing based data center power management in the smart grid environment, *IEEE Communications Surveys & Tutorials*, 16 (1) (2014): 214–233.

9 Enhanced Framework for Energy Conservation and Overcoming Security Threats for Software-Defined Networks

Abhishek Kumar Gaur and Deepak Kumar Sharma

CONTENTS

9.1	Network Security and Security Framework	142
	9.1.1 SDN Support for Security	143
9.2	Possible Threats and Vulnerabilities	144
	9.2.1 Security Issues in SDN Architecture	144
9.3	Reasons of Security Attacks to SDN Architecture	145
	9.3.1 Application Layer Issues	145
	9.3.2 Control Layer Issues	145
	9.3.3 Control Channel	146
	9.3.4 Infrastructure Layer Issues	146
	9.3.5 Faulty System Integration and System Complexity	147
	9.3.6 Security Design and Implementation Inconsistencies	147
9.4	SDN Risk Assessment in Security Framework	147
	9.4.1 Needs of Risk Assessment	148
	9.4.2 Security Assessment	149
	9.4.3 Overall Risk Assessment	149
9.5	Security Enforcement in SDN	151
	9.5.1 Improving SDN Security	152
	9.5.2 Significant Security Research Efforts	152
9.6	SDN Security Solutions	152
	9.6.1 Self-Learning Threat Detection Module	152

DOI: 10.1201/9781003097198-9

9.6.2 NFV/Cloud-Based Security 154
9.6.3 Network State Monitoring and Analysis 155
9.6.4 Integrated SDN Security Framework 156
9.7 Conclusion 156
References 156

9.1 NETWORK SECURITY AND SECURITY FRAMEWORK

Distributed nature of computer networks makes them susceptible to security threats and attacks. Designing the security policies for distributed networks is complex and costly. The centralized nature of SDN provides a global view of the network, which can be used as a means of monitoring and defense. Centralized SDN suffers from the control plane overhead. For addressing and overcoming security challenges, SDN has exploited few existing proposals. SDN popularity has exposed and magnified various security issues or vulnerabilities.

SDN has made the management of cloud services very flexible. For any network application or services, resource requirements and policies are programmed in high-level languages at the application plane. Then these applications are converted into controller understandable programs and passed to the controller using northbound interfaces (REST API). Then controller enforces these policies on the underlying data plane. Attackers may exploit the openness of northbound APIs to attack or create threats to the network. To counter these security issues, it is required to identify, locate the security threats and understand the attack surface. For making SDN secure, a number of security frameworks (StateFit [1], DELTA [2]) have been deployed and evaluated. StateFit follows a centralized architecture for minimizing network latency, overheads and domain-specific state-full security services. Developed solutions detect, prevent and fix security loopholes in terms of network policy updates in a feasible manner. These frameworks promote on-demand security services for SDN. All specified components of the security framework collaborate to deal with given traffic and classify the malicious traffic and involved attack class. Un-optimal security policies sometimes create inconsistency and serious degradation to service quality. Empirical and systematic methods are followed to detect attack scenarios in diverse SDN-based networks targeting the SDN vulnerabilities. SDN employs fuzzy modules for finding unknown security issues. DELTA framework assesses the security of the SDN network. This fuzzy-based framework supports the OpenFlow protocol and OpenFlow controllers.

Security must be treated as a fundamental layer of the computer network. SDN's security framework must be agile, elastic in nature and take care of issues at the data plane, control plane and application plane. Fortinet security framework is to provide security solutions for SDN. Virtual appliances and services are used to provide run-time security at the data plane. Service orchestration frameworks are used for control plane security. Similarly, single pane of glass management is used to provide security at the application plane. The SDN architecture has minimum network complexity and ease of troubleshooting. A preferred approach for SDN deployment is the incremental approach. SDN delay may create network reachability issues and improper enforcement of access control list (ACL) policies. PrePass-Flow [3] is designed as supervised machine learning (logistic regression (LR) and support vector machine (SVM))-based technique to counter data layer failure and ACL violations. It predicts failure, proactively locates, re-computes and installs the ACL policies and predicts for the failed links.

TENNISON [4] is a lightweight framework that is multi-level distributed and does the task of monitoring, resilience, scalability and gives an insight into the effect of multiple controllers on the detection of attacks. Using multiple controllers, the control plane requires coordination among controllers for ensuring consistency.

Case-study-based approach is used to investigate [5] the various facets of cyberattacks on intellectual property. Malicious applications cause potential security threats to the underlying network. A dynamic REST APIs authorization method based on application risk levels can protect against these malicious applications. SEAPP [6] is a REST-based access control framework for network application security. SEAPP manages application permissions, by analyzing permission manifests/byte codes and encrypt REST API calls using NTRU algorithm. SEAPP has a lightweight logic architecture between the application plane and control plane and supports quick deployment and reconfiguration in runtime.

SDN supports the programmable and high-level abstraction of network operations or application development. Most of the security solutions detect and block malicious client, but these solutions have either no or limited provisions on triggering the security fixes or later modifying these fixes. Hence for reinstating the modified fixes at the flagged client requires manually installing the changes or resetting the controller state. Sometimes SDN requires overriding of the previous security fixes with the updated one. The OpenFlow architecture has removed the hardware heterogeneity issues and enabled the security service orchestration on the underlying network. Researchers have proposed a security policy transition framework [7] for flagging malicious clients, traffic reshaping and replacing the previous policies with new security policies.

Designing and implementing integrated security architecture [8] coordinates various security techniques to identify and counter security attacks, on the basis of network policy enforcement. Researchers have designed the integrated security architecture to handle dynamic changes based on the security components. The proposed architecture uses dynamic policy at the SDN controller for enforcing security mechanisms in the switches.

9.1.1 SDN Support for Security

Dynamic flow control replaces the middlebox-based security solution with the software-based solution. These flow rules are generated by the controller that works under the control of network applications. Network applications may make data plane devices act as firewalls, or create access control lists, or traffic redirection. Dynamic flow control using machine learning techniques can differentiate the normal from the suspicious traffic and can instruct the data plane to redirect the suspicious traffic to more specialized security systems.

9.2 POSSIBLE THREATS AND VULNERABILITIES

The logical central nature of the control at the controller may make the single point of failure possible in the SDN. Attackers can exploit the SDN attack surfaces for placing various security threats and causing malicious behavior. All networking device and SDN architecture layers have a number of threats and vulnerabilities attached to them. Further improper configuration and random deployment of these devices make these vulnerabilities even worse to counter.

9.2.1 SECURITY ISSUES IN SDN ARCHITECTURE

Eavesdropping at the control interface is mainly done for tampering with the private and sensitive information transmitted among network appliances or for unauthorized disclosure of critical information. This unauthorized access is done by performing brute-force attacks against administrative credentials or northbound protocols by exploiting network vulnerabilities and then installing infected devices or getting remote connections. Several attacks are intended to mine network control information and forwarding behavior. These attacks broadcast probe packets to selected network devices. Once the network applications are compromised, network policies and internal network control are within the reach of attackers.

Interfaces or protocols (listed in Table 9.1), without the provision of verification/authentication, can be exploited to override or conflict with the existing flow rules. An attacker can generate contradictory policies and enforce them to the network or may change the network topology. Network service disruption can be created by control packet or TCAM flooding. The attacker's created malformed packets can be

TABLE 9.1
Application or Protocol Level Vulnerabilities

Applications/Services/Protocols	Buggy component	Vulnerabilities
OpenSSL	"error state" mechanism, direct call of SSL read()/write()	Confidentiality leakage
RSH	Compatibility with fileio.c	Getting root/admin privileges
SSL	Lack of host name verification in X.509 certificate	man-in-middle attack, communication spoofing
VPN	Management interface of Cisco ASA Software	Random web script injection
Nmap	Android System Server Privilege vulnerability	DoS attack, Arbitrary controller code execution
FTP	Buffer overflow	DoS attack, Random controller code execution, and gaining root permissions
MPLS	Bug in Cisco Carrier Routing System (CRS)	DoS attack, Arbitrary web script injection

inserted into the network for creating improper connection termination or shut-down targeted devices. An attacker can force invalid network topology updates or topology poisoning.

Faulty configurations in network devices, protocols or interfaces result in the form of poor network security policies and when these policies are translated conflicting flow rule are generated. Control or policy information communicated as un-ciphered text across interfaces reveals the network behavior and the attacker, later on, uses them for eavesdropping, malformed packet injection, traffic diversion, changing topology and false device identity.

9.3 REASONS OF SECURITY ATTACKS TO SDN ARCHITECTURE

Attacks at the SDN architecture are primarily done at the application plane, control plane, control channel, or data plane. So we are now discussing the reasons for these attacks on these planes one by one.

9.3.1 APPLICATION LAYER ISSUES

Network policy for the selected SDN is written as the applications and this application lies at the application plane. One should take extra precaution in assigning the authorities or privileges to these policies or applications, while designing or programming. Any extra privilege may cause the abuse of the network infrastructure such as disconnection or shut-down of the network devices or applications.

   a. Service neutralization: Malicious applications can disrupt the network services by manipulating the control packets' contents. These applications can force the controller to discard the control packets, change the order of handling of these packets, modify the service chaining and inspect the control packets for mining control information.
   b. Northbound API passes the compiled network applications to the controller. Misconfigurations of northbound APIs can send the corrected policies to the controller and when the controller enforces these policies on the data plane, it may result in improper termination of a victim application or device or to exposure of network information to unauthorized parties.

9.3.2 CONTROL LAYER ISSUES

In case of conflicts among the flow rules, the SDN controller is not capable of differentiating the new flow rules from the existing flow rules. Keeping this in mind, an attacker can deploy a malicious application for generating rules that conflicts with the existing flow rules and triggers the attacks to the network.

   a. Attacker can also use the protocols and applications security loopholes for poisoning controller and network topology. Host profile vulnerabilities can be exploited for creating LLDP packets such as packets that can be redirected to a fake host in place of the original target.

TABLE 9.2
Attacks at Various Controllers

Unknown Attacks	Traffic Flow	Target Controller
Sequence and Data-Forge	Asymmetric	Floodight
Stats-Payload-Manipulation	Symmetric	Floodligh & OpenDayLight
Echo-Reply-Payload-Manipulation	Symmetric	OpenDayLight
Service-Unregistration	Intra-Controller	OpenDayLight
Link-Discovery-Neutralization	Intra-Controller	FloodLight
Heartbeat-Delay-Randomization	Inter-Controller	ONOS
Missing-Prerequisite	Administrator	FoodLight

b. Compromised applications, interfaces, switches and hosts can trigger the controller's force termination (as in Table 9.2), sensitive information disclosure, redirection of information to the fake location, network policies corruption or unnoted remote connections and might leave the controller to the invalid state.

c. Attackers try to exhaust the controller's/controllers' resources by broadcasting it/them a massive number of fake or fabricated packet-in messages. Hence, all the resources of the controller get wasted.

d. An attacker can force the controller to disconnect the authorized switch by hijacking the genuine switch DPID, or poising the network topology or deploy the threat in the network operating system.

9.3.3 Control Channel

Unencrypted transmission of the control packets at control channels makes it easier for the attacker to get into or mine network control, topology layout and network management information. An attacker then uses ARP poisoning to add an intruder host for launching a man-in-middle attack.

9.3.4 Infrastructure Layer Issues

Attackers can also use ARP poisoning for introducing new unauthenticated controllers that use the original controller identity and force the data plane device to connect to it and disconnect from the original controller; in this way complete network comes under the control of the attackers.

a. Attackers can use the compromised applications/controllers to overwrite or to remove the existing flow rules.

b. Attackers can exploit side-channel, to get the status of how much flow-table is full and then generate the flow-table miss for forwarding packets to the controller and getting new flow rule generation. This process of new flow rule generation continues till the table is overflowed.

Enhanced Framework for Energy Conservation 147

c. A malformed control packet can expose the underlying vulnerabilities to drive the target switch to a risky/undesired state.

d. In side-channel attacks, attackers analyze how a target device response to a given network event and try to get the network's implicit information which can be used for generating other attacks.

9.3.5 FAULTY SYSTEM INTEGRATION AND SYSTEM COMPLEXITY

Most often, an optimal network security system comes with operational complexity and hardness to the users. For the sake of bringing the simplicity of use or reduce complexity, we cannot compromise with security. We analyze the criticality of the application that is going to use the network and on this basis decide the threshold among these parameters.

9.3.6 SECURITY DESIGN AND IMPLEMENTATION INCONSISTENCIES

SDN inconsistency comes to the network because of security solution design or its implementation, unsuitable security solution or missed critical detail/feature of the security solution specification. Following are the cases with the highest probability of inconsistency.

- Security proposals with too many changes
- Adding new devices/protocols in the network
- Threat associated with virtualized services
- Limitation of machine learning-based network solutions
- Side-channel attacks
- Lack of standardization of protocols and interfaces

9.4 SDN RISK ASSESSMENT IN SECURITY FRAMEWORK

Risk assessment of a computer network is done proactively for ensuring the security of flow and traffic routing in SDN. Determining the threat value of SDN component devices against the vulnerability or exposure to these vulnerabilities requires a vulnerability scoring system (CVSS). Secure flow controlling and secure flow rule entries in TCAM of the switches require the accurate risk calculation that is the cumulative threat assessment of SDN devices. This assessment must consider all traffic possibilities in the given network. The SDN controller takes the responsibility of managing or changing the configuration of the network on the best possible flexible, on-demand and dynamic basis. The same thing is done for resource management by the controller [9]. Proper risk assessment extends the SDN system reliability and configuration flexibility against the security loopholes.

Network applications need to maintain an acceptable level of quality of service (as per the service level agreements). Every application has certain specific quality requirements and corresponding parameters. For a given SDN, network policies cannot be static. SDN takes care of enforcing dynamic changing policies and for doing this task SDN controller enforces all required policy changes in networks. SDN protocols make it easier to deal with device heterogeneity and interoperability. One such selected southbound protocol is the OpenFlow [10]. The controller enforces the SDN policies to be reflected as the flow rules in the SDN switches. SDN faces a number of security challenges that must be addressed for efficient utilization of SDN such as

a. Managing complex policies
b. Handling networking delay
c. Absence of standard interfaces/protocols
d. Unavailability of secure flow specification
e. Denial of service
f. Man-in-middle attack
g. Traffic load balancing
h. Traffic reengineering
i. Host-to-host network security
j. Controller and forwarding device security
k. Absence of implicit access control methods
l. Absence of device identity assurance methods and many more

Various security approaches have been established (middlebox, encryption, authentication, Intrusion Detection/Prevention System, etc.) for dealing with some of these raised issues. Among all the above-listed challenges, ensuring the end-to-end security solutions for packets is one of the hardest tasks to handle. Predicting traffic patterns in the current architecture as well as the heterogeneous policy enforcement in distributed policy enforcement environment (multiple controller controlling SDN) creates critical security challenges.

9.4.1 Needs of Risk Assessment

Network application requirements are the key driving forces behind the development of the new network policies. Policies are dynamically converted into network control functions. Policies are converted into the network applications at the application plane; these applications are compiled at the application plane; through the Northbound interface these compiled policies are passed to the controller and then from the controller to the data plane devices. Any change or improper permission assignment to network policies may create security challenges to the whole network. The application plane is exposed to external users and can be easily compromised to perform attacks. Attackers may alter the existing network policies or rules. Unoptimally implemented network functions may cause several attacks to the network such as DoS, hidden tunnel and code injection. Conflicts amongst the flow rules may cause security compromises and poor performance. Further, SDN rules cannot be differentiated on the basis of the time, when these rules are installed. So it may be the case when new rules try to bypass or override the existing rules. Vulnerabilities of

Enhanced Framework for Energy Conservation

the network device (for example various open ports of the switches), end host or malware in generated traffic may be the possible reason for the security threats. Existing SDN security solutions either do not have any provision of traffic pattern and packet contents analysis or have limited provision for the same.

Proper policy enforcement techniques are required for pro-actively handling, analyze the selected traffics against security risks, for strong isolation among network functions and for providing end-to-end security. Common vulnerabilities scoring system (CVSS) [11] is designed that takes threat value, vulnerability and exposure of the networking devices, interfaces and protocols into consideration.

9.4.2 Security Assessment

Most of the security risks in the network are because of the misconfiguration of the network, wrong policy/protocol implementation, absence of handling routines for unexpected events.

SDN must guarantee or have techniques that make it safe from any vulnerability and possible attack. Delta [12] (in Figure 9.1) detects the attack vectors and vulnerabilities of Openflow by employing fuzzy techniques.

9.4.3 Overall Risk Assessment

Risk to the network is the compound effect of the network devices/software weaknesses, degree of an unprotected network device and probable events that expose

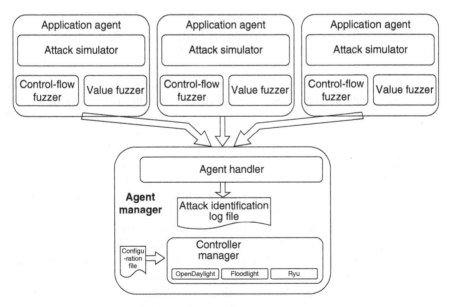

FIGURE 9.1 Delta framework [12].

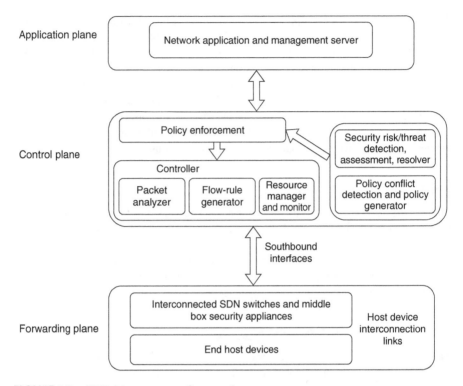

FIGURE 9.2 SDN risk assessment framework.

vulnerabilities (in Figure 9.2). For correct security assessment of network security risks, the following threat [13] models are required:

I. Threats at end-host: Applications (FTP, nmap) that run on end-hosts generate certain kinds of specified traffics. Host vulnerability is the average of vulnerabilities of all running applications. A number of hosts may have security vulnerabilities that can be exposed by attackers. Traffic request's threat value is the product of vulnerability and count of the affected end-host.
II. Threats at switch: Switch's threat value is the average of protocols' (MPLS, LDP) vulnerabilities. The chance of active connections or ports on switches is the exposure to risk. Switch's threat value is the combined effect of switch vulnerability and exposures.
III. Threat at controller: A controller generates flow rules for various network functions/policies. Controller vulnerabilities may result in the generation of conflicting flow rules. The controller's threat exposure can be computed as a ratio of dividing the total dependent-network functions by the number of functions in total. The threat value of a controller comes as the product of controller vulnerability and the degree of controller exposure against these vulnerabilities.

For a given traffic t, overall risk will be the sum of threat values of the hosts, selected access route and network functions involved. The total threat value lies in the range from 0 to 1. Traffic criticality is application-specific and can be high (1), medium (in the range from 0 to 1) or low (0). The security risk of computer networks is proportional to the final calculated threat value and specified traffic criticality. Flow rule generation in SDN is directly controlled by the impact of the risk and risk type. The computational complexity or time to verifying network trust function and consistency of policy check remains constant, while it is variable for conflict resolver.

Network control functions are fundamentally implemented as high-level software applications and used for monitoring and managing the network traffic. These functions are either placed over the top of the controller or as a separate data consumer function. SDN security is one of the most sensitive issues. SDN suffers from two classes of security problems: one class (says class A) contains the problem that is the same as the conventional network problems and the other class (says class B) contains the problem that is due to the weakness of poor SDN architecture, protocols and interfaces. Class A problems can be countered by developing SDN security policies. Later on, these security policies are converted into high-level application programs. Further, these applications are compiled and passed to the controller. Controller enforces them in a timely and reliable manner, over forwarding device. Forwarding devices in the SDN do use the flow rules stored in their TCAM memory for handling the application traffic. Class B problems require the security modification in the SDN architecture level so that changes of attacking the network device can be reduced to zero.

Security applications for SDN replace the hardware-based security middle-boxes with the programmable switches that utilize the machine learning and virtualization techniques assisted security rules. Machine learning techniques also help in threat detection. SDN architecture does not mature enough against security risks and threats, present at various interfaces and at network elements level. So for making the SDN optimal secure, the need for a reference architecture with embedded security arises. SDN can be fully utilized, only after it is proved to strictly fulfill the minimum set of optimal security standards.

9.5 SECURITY ENFORCEMENT IN SDN

Risk analyze framework [14] uses network policy conflict resolver and consistency checker for ensuring end-to-end security. Network trust function verifies the dynamically generated policies to check the certificate validity, domain name mapping and credibility of the certificate issuing authority. The trust function alerts against network security violations.

Policy conflict resolver finds and overrides conflicts of generated flow rules, corresponding to two or more currently active policies in a given network. Conflict resolver uses pattern matching of conflict detection and artificial intelligence rule overriding. While overriding, it takes the assistance of rules prioritizing levels as defined by network administrators. Policy consistency check makes the network free from any kind of inconsistency in policies specified at the application plane and that of enforced or reflected at the data plane. The mismatch between application plane policies and flow rules implemented at the data plane may make the network unstable and conflict-prone. Network bad configurations, protocol vulnerabilities and various device-level security loopholes are used for extracting the hidden threat or derive attack models for given network traffic. The overall risk for selected traffic is calculated as the weighted sum of the threat values and traffic criticality. Flow rules of the traffic are now generated by taking count of this calculated overall risk.

9.5.1 Improving SDN Security

The SDN framework has new interfaces and protocols in comparison to tradition networks. These newly added entities may introduce the new kind of threats. For securing the communication interfaces and data interfaces must exchange the data/control information in encrypted packets. SDN devices must be authenticated and check against the pre-existing trust mechanism, no matter whether the device is added at the control or data plane. Controller must have the ways to authorize not only the trusted devices but also network applications in the form of certificate signed by some trust authority. Ensuring and cross-verifying network trusted guarantees that most of the malicious entities will be isolated from SDN.

SDN must deal with the conflicting flow rules that may be either generated due to conflicts among applications or with security policies. SDN must have a mediator and policy checking approach for resolving these conflicts. DoS, topology poisoning and fake device effect the network state. SDN must have robust framework network state monitoring (warning, risks, stable states) for keeping the network safe. Security framework comes with tools for secure network applications and policies. SDN applications are exhaustively tested against the functional, non-functional and coverage test before being deployed in the network.

9.5.2 Significant Security Research Efforts

Researchers are exploiting SDN to locate the possible areas for security enhancements (as given in Table 9.3). One of the areas for such improvement is to handle the threats that are present in the system due to the SDN's centralized nature and programmability [32].

9.6 SDN SECURITY SOLUTIONS

SDN security solutions come from the application of machine learning, network function virtualization and applications of the cloud.

9.6.1 Self-Learning Threat Detection Module

For the detection of threats, SDN utilizes machine learning techniques (especially supervised learning). For the online generation of the training data set for the supervised learning model, the SDN controller selects and labels some port and

TABLE 9.3
Significant Security Research Proposals

S.No.	Proposed Security Solution	Merits or Benefits
1	OpenSAFE [15]	Deals with arbitrary traffic using ALARMS (flow specification language)
2	FRESCO [16]	Modular-based security services development based on threat detection and flow monitoring
3	FleXam [17]	Locate and reduces security threats by analyzing packet level details
4	NICE [18]	Uses attack graphs, does intrusion detection for securing Openflow applications
5	FlowChecker [19]	Finds policy level inconsistences in multi-controller SDN
6	FLOVER [20]	Finds and deals with conflict in flow rules for various network policies
7	FortNOX [21]	Use role-based authentication and signature-based authentications for fixing flow rule entry installation conflicts
8	Identity-Based Cryptography Protocol [22]	Ensures all communication interfaces except northbound, for distributed SDN
9	Secure NBI for SDN applications with NTRU	NTRU and NSS signature-based sharing of network details with authorized network programs
10	Attribute-based encryption framework [23]	Ensure access control methods in SDN
11	Intelligent Risk Assessment Based on Neural Networks [24]	Deals with DDoS attacks in SDN
12	HiFIND [25]	Protects SDN users and network service providers for DDoS attacks
13	SN-SECA [26]	Integrates and validates various security parameters for secure SDN
14	SECO [27]	Detects and secures control against host or network device originated DoS
15	Securing distributed SDN controller [28]	Considers mapping algorithms, synchronization / heartbeat messages to protect the network from DoS
16	STRIDE [29]	Secures the southbound interface against possible threats in communication
17	DAC [30]	Ensure secure dynamic access control on network interfaces on network applications
18	SIMPLE-fying Middlebox Policy Enforcement [31]	Handles overhead of middlebox-based security for SDN, without any change to middlebox

protocols, and collects data about the network state and traffic patterns. The threat detection module to SDN is developed as the controller application and once it is successfully trained, it intelligently detects threats (SD-TDPS).

A trained threat detection module is a set of machine learning features to perform the packet inspection and detection of attack type/sources. The detection module labels the features that differentiate the normal from the anomaly traffic. For the optimal use of the network computational resources, SDN IDS limits the features' reporting task to edge-switches and this reporting repeatedly happens after a fixed interval of time.

The NIDS architecture [33] takes the benefits of either a signature-based security approach or machine learning with a back-propagation approach for the recognition of attacks' patterns.

a. Signature approach works for the known type of attacks and comes as a physical device attached to the network. Machine learning-based attack detector works for both known and unknown types of attacks. In this system, the controller collects the traffic statistics, uses them for training, detection of anomalies, and instruct the controller to generate new flow entries.
b. Security framework named Athena [34] is deployed for distributed SDN controllers. This framework has separate blocks, for feature selection, attack detection, security solution generation, resource management and user interface management. Athena supports the automatic development of threat detectors.

9.6.2 NFV/Cloud-Based Security

Cloud-based security solutions reduce the controller load and transfer the security function (firewall, IDS) processing to a cloud environment. An NFV-based security solution deploys the security functions as virtualized functions. Hybrid stateful firewall function [35] provided security functions are again NFV based, but these NFV are implemented on the cloud. A security architecture [36] is developed to convert the security policies into virtualized services. Security policies are mapped to security function as a combination of separation, chaining, merging and recording functions.

Cloud-based security functions are integrated into data plane filtering [37], for a better security architecture. Techniques of selection and enforcement of the cloud-based heterogeneous network security function are developed in terms of standardized interfaces. This framework also decides when to analyze the packet at data plane devices (lightweight treatment), and when to send it to the cloud environment (heavyweight treatment) for inspection and attack detection. Further, a security framework [38] is developed as a result of combining security meta-functions database and service orchestration model that translates the high-level security requirements into NFV. Later on, these virtualized functions are implemented in the data plane.

- The SDN controller always keeps the network current state and to counter the attacks at the network generates or modifies the flow entries. For handling the link congestion (DoS), a policy management and enforcement framework [39]

is developed that stores the high-level network policies in the policy database and then uses them to calculate the congestion-free paths.
- Flow obfuscation is used to counter the side-channel attack. This technique utilizes the packet's header field contents modification utility of the OpenFlow architecture. For counter probe flow, the controller generates a single flow rule for the switch to update certain packet header fields, then forward it to the next switch, and this rule flow rule generation for field updating and forwarding is repeated for a fixed number of switches. Finally, the security policy is enforced by the controller by issuing corresponding flow rules. For resolving flow table overflow, a flow rule can be generated using optimal routing aggregation algorithms, and a multi flow-table architecture is used for extending the flow table capacity.
- Using passive or active inspection, Network Flow Guard (NFG) [40] can detect unnoticed or unsafe access points. In passive inspection, we first allow the suspect device to connect to the network, looks the Time-to-live for the packets to track the presence of unauthorized access points, then MAC or IP addresses linked to per port of switches is counted; it should be one per port. Port with a count of more than one are flagged and that switch is separated from the network. Finally, the average TTA (time-to-acknowledge) for TCP SYN messages for every port is calculated. If any port average TTA is observed greater than the global average the port is flagged and its packet flow is redirected to the active detection engine.
- Packet Checker [41] collects the desired header field values for the first Packet_In packet coming from every switch connected to the current network and stores these into hash maps as network-information-base. Same fields of other subsequent coming Packet_In messages from these switches are matches with the previously recorded values to drop the malformed packet injections.
- Device fingerprinting-based approach [42] is developed to authenticate the devices that are trying to connect to the network. If the newly added device failed to authenticate its fingerprint, the fingerprint module alert about this unverified device. Secure access control [43] in SDN enforces the IEEE802.1x standard and the EAP (Extensible Authentication Protocol) along with RADIUS server implementation.
- Centralized nature of SDN is used for creating secure sessions [44] by validation features against a set of constraints selected from the existing security policies. KISS [45] does the device registration and association task for establishing a secure channel. KISS uses a random key generator to work on a per-packet basis. In this way, KISS has resolved the TLS issue.
- Third-party applications face access level administrative permission issues. SM-ONOS [46] manages the access level permissions for network controllers and over devices. SM-ONOS enforces application access level policies as set by the developer by role basis and access policies due to manual constraints.

9.6.3 Network State Monitoring and Analysis

The attacks change the state of the SDN. Network monitoring identifies these state-changes and once these state-changes are detected, countermeasures are taken to bring the network back to a secure state.

- In Path Class Approach (PCA) [47], data objects are mapped to certain secure forwarding paths. The selected path provides protection against certain security requirements and security policies. In this way, SDN ensures network confidentiality, data integrity and maximum availability. PCA also protects the system against DoS attacks by reducing path congestion.
- Security plane is introduced [48] to making the SDN more secure and reducing the security processing load from the controller. This plane has security agents that are distributed in the data plane and attack detection engine to work in parallel to the controller.
- Sphinx [49] is developed to provide run-time flow anomalies detection. Sphinx is placed as a layer in between the network control plane and the data plane. Sphinx collects the meta-data from the OpenFlow packets and saves the network from the network topology and invalid states and has information of the entire network topology and forwarding paths.

9.6.4 INTEGRATED SDN SECURITY FRAMEWORK

SDN's security framework is required to have fine-grained control at all security functions and also have the most efficient security components to solve specific issues. To cover a wide range of scenarios, the framework must deal with high-level features, in place of low-level details. To counter network attacks, a security architecture framework [50] is developed that has fine-grained overflow entries generation/updation, scans packets for detecting threats and authorizes network applications.

9.7 CONCLUSION

Energy is considered one of the most critical points of concern for every network designer or engineer. A network can optimally utilize the networking resources only if it is free from any kind of security attacks or threats like DoS and other possible vulnerabilities. These threats cause poor network utilization. The presented work has covered the listing of the research works that have targeted the discovery and overcoming of attacks and vulnerabilities to the SDN. Research works on various risks and their assessments and discovered security solutions are also included.

REFERENCES

[1] R. Hwang, V. Nguyen, P. Lin, "StateFit: a security framework for SDN programmable data plane model", *2018 15th International Symposium on Pervasive Systems, Algorithms and Networks (I-SPAN)*, Yichang, China, 2018, pp. 168–173, doi:10.1109/I-SPAN.2018.00035.

[2] S. Lee, J. Kim, S. Woo, C. Yoon, S. Scott-Hayward, V. Yegneswaran, et al., "A comprehensive security assessment framework for software-defined networks", *Computers & Security*, 91 (April) (2020): 101720.

[3] M. Ibrar, L. Wang, G.-M. Muntean, A. Akbar, N. Shah, K.R. Malik, "PrePass-flow: a machine learning based technique to minimize ACL policy violation due to links failure in hybrid SDN", *Computer Networks*, 184(15) (January) (2021): 107706, doi:10.1016/j.comnet.2020.107706.

[4] L. Fawcett, S. Scott-Hayward, M. Broadbent, A. Wright, N. Race, "TENNISON: a distributed SDN framework for scalable network security", *IEEE Journal on Selected Areas in Communications*, 99 (September) (2018): 1–1, doi:10.1109/JSAC.2018.2871313.

[5] S. Banyal, Amartya, D. Kumar Sharma, "Security vulnerabilities, challenges, and schemes in IoT-enabled technologies", *Blockchain Technology for Data Privacy Management*, CRC Press, 1st Edition, 2021, p. 28, ISBN: 9781003133391.

[6] T. Hua, Z. Zhang, P. Yi, D. Liang, Z. Li, Q. Ren, et al., "SEAPP: a secure application management framework based on REST API access control in SDN-enabled cloud environment", *Journal of Parallel and Distributed Computing*, 147, (January) (2021): 108–123, doi:10.1016/j.jpdc.2020.09.006.

[7] J.H. Cox, R.J. Clark, H.L. Owen, "Security policy transition framework for software defined networks", *2016 IEEE Conference on Network Function Virtualization and Software Defined Networks (NFV-SDN)*, Palo Alto, CA, Nov., 2016, pp. 7–10, doi:10.1109/NFV-SDN.2016.7919476.

[8] K.K. Karmakar, V. Varadharajan, U. Tupakula, M. Hitchens, "Towards a dynamic policy enhanced integrated security architecture for SDN infrastructure", *NOMS 2020—2020 IEEE/IFIP Network Operations and Management Symposium*, Budapest, Hungary, April, 2020, pp. 20–24, doi:10.1109/NOMS47738.2020.9110405.

[9] Ghodsi et al., "Intelligent design enables architectural evolution", *Proceedings of 10th ACM Workshop Hot Topics in Networks*, Article No. 3, 2011, pp. 1–6.

[10] OpenFlow White Paper, [Online]. Available: http://archive.OpenFlow.org/

[11] P. Mell, K. Scarfone, S. Romanosky, "Common vulnerability scoring system, security & privacy", *IEEE*, 4 (6), (December) (2006): 85–89.

[12] S. Lee, C. Yoon, C. Lee, S. Shin, V. Yegneswaran, P. Porras, "Delta: a security assessment framework for software-defined networks", *Proceedings of the 2017 Network and Distributed System Security (NDSS) Symposium*, 17 (2017).

[13] Khera, A., D. Singh, D.K. Sharma. "Information security and privacy in healthcare records: threat analysis, classification, and solutions", in *Security and Privacy of Electronic Healthcare Records: Concepts, Paradigms and Solutions*, (2019), p. 223.

[14] B.K. Tripathy et al., "A novel secure and efficient policy management framework for software defined network", *IEEE 40th Annual Computer Software and Applications Conference*, IEEE, 2016, pp. 423–430.

[15] J.R. Ballard, I. Rae, A. Akella, "Extensible and scalable network monitoring using opensafe", *Proc. INM/WREN*, 2010.

[16] S. Shin et al., "FRESCO: modular composable security services for software-defined networks", *Proceedings of Network and Distributed Security Symposium*, 2013.

[17] S. Shirali-Shahreza, Y. Ganjali, "Efficient implementation of security applications in OpenFlow controller with FleXam", *21st IEEE Annual Symposium on High-Performance Interconnects*, IEEE, 2013, pp. 49–54.

[18] M. Canini, et al., "A NICE way to test OpenFlow applications", *Proceedings of the 9th USENIX conference on Networked Systems Design and Implementation*, 2012.

[19] E. Al-Shaer, S. Al-Haj., "Flowchecker: configuration analysis and verification of federated OpenFlow infrastructures", *Proceedings of the 3rd ACM workshop on Assurable and Usable Security Configuration*, 2010.

[20] S. Son et al., "Model Checking Invariant Security Properties in OpenFlow", http://faculty.cse.tamu.edu/guofei/paper/Flover-ICC13.pdf.

[21] P. Porras et al., "A security enforcement kernel for OpenFlow networks", *Proceedings of the First Workshop on Hot Topics in Software Defined Networks*, ACM, 2012, pp. 121–126.

[22] J.-H. Lam, S.-G. Lee, H.-J. Lee, Y.E. Oktian, "Securing distributed SDN with IBC", *Seventh International Conference on Ubiquitous and Future Networks*, Sapporo, 2015, pp. 921–925.

[23] Y. Shi, F. Dai, Z. Ye, "An enhanced security framework of software defined network based on attribute-based encryption", *4th International Conference on Systems and Informatics (ICSAI)*, Hangzhou, 2017, pp. 965–969.

[24] Gabriel et al., "Achieving DDoS resiliency in a software defined network by intelligent risk assessment based on neural networks and danger theory", *15th IEEE International Symposium on Computational Intelligence and Informatics*, 2014, pp. 319–324.

[25] Z. Li, Y. Gao, Y. Chen, "HiFIND: A high-speed flow-level intrusion detection approach with DoS resiliency", *Computer Network*, 54 (8), (2010): 1282–1299.

[26] D.V. Bernardo, B.B. Chua, "Introduction and analysis of SDN and NFV security architecture (SN-SECA)", *2015 IEEE 29th International Conference on Advanced Information Networking and Applications*, Gwangiu, 2015, pp. 796–801.

[27] S. Wang, K.G. Chavez, S. Kandeepan, "SECO: SDN sEcure COntroller algorithm for detecting and defending denial of service attacks", *2017 5th International Conference on Information and Communication Technology (ICoIC7)*, Melaka, 2017, pp. 1–6.

[28] W. Etaiwi, M. Biltawi, S. Almajali, "Securing distributed SDN controllers against DoS attacks", *2017 International Conference on New Trends in Computing Sciences (ICTCS)*, Amman, 2017, pp. 203–206.

[29] H. Shawn, L. Scott, O. Tomasz, S. Adam, "Uncover Security Design Flaws Using The STRIDE Approach, March 2015, http://msdn.microsoft.com/engb/magazine/cc163519.aspx

[30] Y. Tseng, M. Pattaranantakul, R. He, Z. Zhang, F. Nat-Abdesselam, "Controller DAC: securing SDN controller with dynamic access control", *2017 IEEE International Conference on Communications (ICC)*, Paris, 2017, pp. 1–6.

[31] Z.A. Qazi, C.-C. Tu, L. Chiang, R. Miao, V. Sekar, M. Yu, "SIMPLEfying middlebox policy enforcement using SDN", *ACM SIGCOMM*, August 2013.

[32] D. Kreutz, F. Ramos, P. Verissimo, "Towards secure and dependable software-defined networks", *Proceedings of the Second ACM SIGCOMM Workshop on Hot Topics in software Defined Networking*, 2013, pp. 55–60.

[33] A. Pranggono, "Machine learning based intrusion detection system for software defined networks", *2017 Seventh International Conference on Emerging Security Technologies*, EST, 2017, pp. 138–143, doi:10.1109/EST. 2017.8090413.

[34] S. Lee, J. Kim, S. Shin, P. Porras, V. Yegneswaran, "Athena: a framework for scalable anomaly detection in software-defined networks", *47th Annual IEEE/IFIP International Conference on Dependable Systems and Networks*, DSN, pp. 249–260, doi:10.1109/DSN.2017.42.

[35] C. Lorenz, D. Hock, J. Scherer, R. Durner, W. Kellerer, S. Gebert, et al., "An sdn/nfv-enabled enterprise network architecture offering fine-grained security policy enforcement", *IEEE Communication Magazine*, 55 (3) (2017): 217–223, doi:10.1109/MCOM.2017.1600414CM.

[36] W. Lee, N. Kim, "Security policy scheme for an efficient security architecture in software-defined networking", *Information*, 8 (2) (2017): 65.

[37] S. Hyun, J. Kim, H. Kim, J. Jeong, S. Hares, L. Dunbar, et al., "Interface to network security functions for cloud-based security services", *IEEE Communication Magazine*, 56 (1) (2018): 171–178, doi:10.1109/MCOM.2018.1700662.

[38] Z. Lin, D. Tao, Z. Wang "Dynamic construction scheme for virtualization security service in software-defined networks", *Sensors* 17 (4) (2017): 920.

[39] R. Sahay, G. Blanc, Z. Zhang, K. Toumi, H. Debar, "Adaptive policy-driven attack mitigation in SDN", *Proceedings of the 1st International Workshop on Security and Dependability of Multi-Domain Infrastructures*, ACM, 2017, p. 4.

[40] J.H. Cox, R. Clark, H. Owen, "Leveraging SDN and webRTC for rogue access point security", *IEEE Transactions Network Service Management*, 14 (3) (2017): 756–770, doi: 10.1109/TNSM.2017.2710623.

[41] S. Deng, X. Gao, Z. Lu, X. Gao, "Packet injection attack and its defense in software-defined networks", *IEEE Transactions on Information Forensics and Security*, 13 (3): 695–705, doi:10.1109/TIFS.2017.2765506.

[42] N. Gray, T. Zinner, P. Tran-Gia, "Enhancing SDN security by device fingerprinting" *IFIP/IEEE Symposium on Integrated Network and Service Management, IM*, 2017, pp. 879–880, doi:10.23919/INM.2017.7987393.

[43] D.M.F. Mattos, O.C.M.B. Duarte, "Authflow: authentication and access control mechanism for software defined networking", *Annals of Telecommunication*, 71 (1112): 607–615.

[44] A. Ranjbar, M. Komu, P. Salmela, T. Aura, "An SDN-based approach to enhance the end-to-end security: SSL/TLS case study", *NOMS 2016—2016 IEEE/IFIP Network Operations and Management Symposium*, 2016, pp. 281–288, doi:10.1109/NOMS.2016.7502823.

[45] D. Kreutz, P.J.E Verssimo, C. Magalhaes, F.M.V. Ramos, "The kiss principle in software-defined networking: a framework for secure communications", *IEEE Security & Privacy*, 16 (5): 60–70, doi:10.1109/MSP.2018.3761717.

[46] C. Yoon, S. Shin, P. Porras, V. Yegneswaran, H. Kang, M. Fong, et al., "A security-mode for carrier-grade SDN controllers", *Proceedings of the 33rd Annual Computer Security Applications Conference*, ACM, pp. 461–473.

[47] K. Wrona, S. Oudkerk, S. Szwaczyk, M. Amanowicz, "Content-based security and protected core networking with software-defined networks", *IEEE Communication Magazine*, 54 (10): 138–144, doi:10.1109/MCOM.2016.7588283.

[48] A. Hussein, I.H. Elhajj, A. Chehab, A. Kayssi, "SDN security plane: an architecture for resilient security services", *IEEE International Conference on Cloud Engineering Workshop (IC2EW)*, pp. 54–59, doi:10.1109/IC2EW.2016.15.

[49] M. Dhawan, R. Poddar, K. Mahajan, V. Mann, "Sphinx: detecting security attacks in software-defined networks", *Proceedings of the 2015 Network and Distributed System Security (NDSS) Symposium*, 2015.

[50] Z. Hu, M. Wang, X. Yan, Y. Yin, Z. Luo, "A comprehensive security architecture for SDN", *18th International Conference on Intelligence in Next Generation Networks*, 2015, pp. 30–37,doi:10.1109/ICIN.2015.7073803.

10 Smart Shopping Trolleys for Secure and Decentralized Services

Jaspreet Singh, Charanjeet Singh, Yuvraj Singh, Chamandeep Singh and Monica Bhutani

CONTENTS

10.1 Introduction 161
    10.1.1 Background 162
        10.1.1.1 IoT 162
        10.1.1.2 YOLO 162
        10.1.1.3 Blockchain Technology 162
10.2 Proposed Architecture 163
    10.2.1 Overview 163
    10.2.2 Implementation 163
    10.2.3 Rationale 164
    10.2.4 Data Sharing 164
    10.2.5 Use Cases 166
10.3 Scenarios Study 166
    10.3.1 Current Scenario 167
    10.3.2 Solution by EasyBills 167
    10.3.3 Results 167
10.4 Conclusion 167
References 168

10.1 INTRODUCTION

Nowadays, the proliferation of supermarkets and shopping malls adds to the rapid development of Internet of Things (IoT) technology. The technology has produced various intelligent systems for helping customers in shopping efficiency. Some specific outcomes are using the transport for products by utilizing a mobile shopping platform and improving the information provided to the user more effectively.

    The IoT technology sets a network of sensors and actuators in the trolley that can function in data collection and customer management. It includes functions like

DOI: 10.1201/9781003097198-10

checking the proximity of items, their RFIDs, objects and having them in the cart. It significantly improves the quality of the shopping service. This chapter proposes a cost-effective way to keep check on thefts in markets and establish a user-interactive scheme with decentralized services. It aims to reduce the workforce and promote the shopping experience for its customers. For the ease of customers, wireless technology provides robustness and ease. Finally, experiments are conducted in a familiar environment to present the encouraging results for real-world deployment.

10.1.1 BACKGROUND

This section includes the background knowledge required for the tech stack used in this chapter. Various domains are interconnected to make the shopping experience easy and secure for customers.

10.1.1.1 IoT

The IoT means a network of objects that are connected through various sensors, embedded systems and radio frequency identification systems to communicate and exchange the data for performing some specific task.

Recently, an enormous amount of advancements in this field has given way to new applications and fields. Interfacing with sensors and actuators plays the combined role of environmental sensing, specific computing and wireless communication. Factors accompanying the effectiveness of miniaturization of hardware, fast sensing equipment, energy saving and scavenging and the fact that many applications cannot be wired make IoT technology suitable for various application domains such as medicine and health care, environment and industrial monitoring.

For example, a flexible bill shopping system can be created by connecting all the items in a grocery store; an RFID tag would be attached to every object which would help it get recognized by an RFID reader. Then data is to be collected and transferred to a central server. The benefits of using such an IoT-based application include a smooth, faster and secure billing facility for customers.

10.1.1.2 YOLO

YOLO is a convolutional neural networks family that achieves near state-of-the-art results with a single end-to-end model that can perform object detection in real time.

YOLO involves a single deep convolutional neural network (initially a version of GoogLeNet, later updated and called DarkNet based on VGG) that splits the input into a grid of cells and each cell directly predicts a bounding box and object classification. Smart shopping trolleys have cameras with applied YOLO to the full image of an object. This network divides the image into regions and predicts bounding boxes and probabilities for each part. Bounding box weight represents the predicted probabilities.

10.1.1.3 Blockchain Technology

With blockchain technology, there comes the power of transparency in the system. It is a secure decentralized system where dependent parties can collaborate. There is

Smart Shopping Trolleys

proof for every transaction happening in the system. Hence, it can be of utilization where trust is required.

One of the most-used cases of blockchain technology is in the supply chain of products. Products reach supermarkets via various intermediaries. The utmost requirement of every user is to purchase a high-quality product. However, tracking it through this chain can be difficult. To track each product's history, blockchain technology provides a transparent mechanism for customers to verify the product's quality parameters.

10.2 PROPOSED ARCHITECTURE

The architecture of the working of the trolley, its implementation and rationale is described in this section.

10.2.1 OVERVIEW

The proposed architecture involves a microcontroller integrated with a camera, infrared sensors, RFID scanners, LCD screen and weighing machine. RFID scanners are present at the front of the trolley. It scans the RFIDs of the product and adds them to the cart. A camera that is placed just above the shopping trolley captures every item in the trolley. Infrared sensors and cameras assure that an RFID reader scans all the things inside. The LCD screen is present at the place where the customer holds the trolley. LCD screen shows the items added to the cart, their prices, supply histories and the options to pay online. It also shows the virtual model of the mall and guides the customer who wants to find any specific item.

10.2.2 IMPLEMENTATION

Each trolley is installed with IoT sensors like proximity sensors, RFID readers, memory card linked with LCD screen having the capability to store the customer's cart and online payments. The microcontroller is programmed such that when an item is in the proximity of the trolley, the LCD screen asks for placing it before an RFID reader. Each item that has RFID has data in it regarding the item details. The item is added to the cart when brought near the RFID reader. The object detection algorithm is used for recognizing items like vegetables or fruits without RFID tags. The items are placed on the weighing machine. The weight is shown on the screen and added to the cart. We are using TensorFlow APIs for object detection, an open-source framework built on top of TensorFlow that makes it easy to construct, train and deploy object detection models. The TensorFlow Object Detection API also uses Protobufs to configure model and training parameters. Before the framework can be used, the Protobuf libraries were compiled. Libraries that were used include Protobuf 3.0.0, Python-tk, Pillow 1.0, Lxml, tf Slim, Matplotlib, Tensorflow (>=1.12.0), Cython and cocoapi. The object detection model is run locally on pre-trained models of YOLO for objects.

If any customer wants to find any item, he/she can search it on the screen. The guide to the destination is shown. The indoor navigation is realized by Unity3D using Google ARCore. Motion tracking and real-time navigation are provided for customers within the mall for all the items.

If any customer wants product details for quality assurance, he/she can check product history from the screen. The product scanned supply chain can be shown on the screen with the proof of transparency and secure database of blockchain. Customers can book for items with the mall sitting from home. He/she can pick up the prepaid order.

10.2.3 RATIONALE

The rationale for a customer entering the mall is provided step by step in the following points. Figure 10.1 shows the flowchart for the system.

1. *Self check-in:* The person entering the mall has to go to collect the trolley where each trolley has a unique ID. This is a mandatory process for anyone who wants to do the shopping. For persons coming with family, only one person is needed to have the trolley.
2. *Customer's cart:* The cart of the user is maintained as the user picks and adds an item in the trolley. For items with RFIDs, they are required to be scanned before placing inside the trolley. For items without them like fruits, the weighing machine is present that weighs the item and adds them to the cart using an LCD screen.
3. *Sensors security:* The camera, infrared sensors and the UHF RFID long-range reader ensure that there is no chance that a person picks the item in the pocket or tries to steal.
4. *Guide and product history:* Customers are provided with each product's location in the mall using in-store maps. They can quickly find the items they want and save time. The quality assurance of each tagged product is also provided to customers using blockchain-powered supply chain history.
5. *Billing and checkout:* Customers can pay online for the items in their trolley when they place back the trolley for billing. They can then easily leave the mall. So, there is no requirement to check the bill by guards at exit gates, thus saving time.

10.2.4 DATA SHARING

The data-sharing mechanism used between customers and the shops can risk customers' data privacy if it is transparent between the group of companies. The data-sharing method will be based on customers' consent for any mall. Hence, the customer will have availability to every mall and pre-book the order and collect once he/she permits to share information. It allows customers to efficiently pick their paid orders from any mall in the chain.

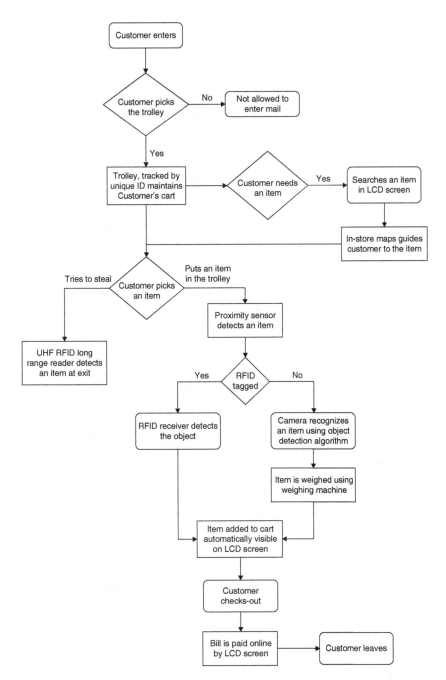

FIGURE 10.1 Implementation flow.

10.2.5 USE CASES

Different use cases can exist for the architecture. Some of them are listed below:

- *B2B Commerce:* selling, buying and trading of goods and services through an online sales portal between businesses.
- *Business Process Engineering and Design:* focus on the analysis and design of workflows and business processes.
- *Business Process Outsourcing (BPO):* method of subcontracting various business-related operations to third-party vendors.
- *Contracting in Supply Chains:* includes revenue-sharing contracts, return contracts, two-part tariff contracts, service commitment contracts, quantity flexibility contracts, etc.
- *E-Commerce and E-Business Models:* when a company markets its products or services directly to other businesses, for example eBay.com.
- *Electronic Markets, Auctions and Exchanges:* a virtual online market to conduct, business-to-business e-commerce over the internet.
- *Enterprise Management Systems:* enterprise-scale application software which addresses customer's software needs.
- *Queuing Networks:* systems in which single queues are connected to routing network.
- *Service Logistics and Product Support:* Advising and guiding customers on the shipping process.
- *Supply Chain Management:* management of the flow of goods and services and includes all processes that transform raw materials into final products.
- *Technology, Information and Operations*: include automotive industries, electronics manufacturing, financial services, healthcare, industrial equipment, media, entertainment, process industries, retailing and telecommunications.

10.3 SCENARIOS STUDY

This section highlights the key differences in the improvement of the current scenarios. The following table highlights the main points.

Current Scenario	Solution
The queue at billing counters	The billing is possible directly by the trolley
Theft detection is not possible real time	Theft detection is possible real time with the sensors
Need for mobile app to automate cart like Amazon Go	No need for mobile apps pre-installed, the trolley works as an integrated system
Need for guards to verify bills	Auto-verified billing
Wrong billings and human errors	No chance of human error
Lack of navigation for each item	Indoor navigation for every item is possible
Each mall works standalone in a centralized manner. Billing possible only at that mall	A decentralized system of companies allows pay and pick up from any mall facility

10.3.1 CURRENT SCENARIO

Billing counters suffer from long lines that make the customers impatient in the billing counter. Much time is wasted just for the billing process and verification. Amazon Go has automated carts and requires their customers to install the mobile app. It is not suitable for everyone. Bills are generally verified with the items one is carrying by the guards during the exit. Incorrect billings and human errors are possible. Hence, the lack of proper verification is common. Lack of navigation for each item is not present; customers need to search for items in the mall. There is no guidance for him/her to find all the required items easily. User experience suffers in such a scenario. Also, each mall works as a standalone in a centralized manner. Billing is possible only at that mall. One requires items from the mall but cannot go inside the mall due to time constraints or any pandemic situation. In those times, there is a need for decentralized trust.

10.3.2 SOLUTION BY EASYBILLS

Billing is possible directly by the trolley. It saves much of the time wasted on the billing process and verification. The user interface provides the object recognition function and assistive information to improve the shopping experience for customers. There is no need for mobile apps to be pre-installed. The trolley works as an integrated system. Billing is auto verified and there is no chance of human error. Indoor navigation for every item is possible. A decentralized network of companies allows pay and pick up from any mall facility, thus saving time and ensuring security. Finally, through the recorded data that links customers and shopping behavior, the proposed smart shopping trolley demonstrates the high potential capability to integrate into the IoT in supermarkets and malls.

In times of health emergencies like during COVID-19, this process can be beneficial as it involves the contactless buying and billing of items.

10.3.3 RESULTS

The system of sensors employed in the trolley works with a delay of a few milliseconds. The Micro SD card can store the operational data of the customer's cart. The YOLO model used for object detection has several advantages over classifier-based systems. It looks at the whole image and predictions that are informed by the global context. It also makes predictions with a single network evaluation, unlike other systems like R-CNN that require thousands for an available image. This makes it extremely fast. It is more than 1000x faster than R-CNN and nearly 100x quicker than fast R-CNN.

10.4 CONCLUSION

This chapter develops a smart shopping cart for use in supermarkets and malls. The user interface provides the facial recognition function and assistive information to improve the shopping experience for customers. For a better shopping experience, the automatic billing facility will help customers avoid queues during the checkout process. Finally, through the recorded data which links between customers and shopping behavior, the proposed smart shopping trolley demonstrates the high potential capability to integrate into the IoT in supermarkets and malls.

REFERENCES

[1] S. Ren, K. He, R. Girshick, J. Sum, "Faster R-CNN", Published at the NIPS (Neural Information Processing Systems Conference), 2015.
[2] C. Dicle, M. Sznaier, O. Camps, "The way they move: tracking multiple targets with similar appearance", IEEE International Conference on Computer Vision, 2013.
[3] E. Maggio, M. Taj, A. Cavallaro, "Efficient multi-target visual tracking using random finite sets", *IEEE Transactions on Circuits and Systems for Video Technology*, 18 (8), (Aug.) (2008): 1016–1027.
[4] L. Leal-Taixé, C. Canton-Ferrer, K. Schindler, "Learning by tracking: siamese CNN for robust target association", Proceedings of the IEEE Conference On Computer Vision Pattern Recognition Workshops (CVPRW), Jun./Jul., 2016, pp. 418–425.
[5] R. Padilla, C.F.F. Costa Filho, M.G.F Costa "Evaluation of HAAR cascade classifiers designed for face detection", *World Academy of Science, Engineering and Technology*, 64 (2012): 362–365.
[6] W. Zhiqiang, L. Jun, "A review of object detection based on convolutional neural network (CNN)", 26 (2013): 2553–2561.
[7] A. Bewley, Z. Ge, L. Ott, F. Ramos, B. Upcroft, "Simple online and realtime tracking", IEEE International Conference on Image Processing (ICIP), 2016, pp. 3464–3468.
[8] Z.W. Pylyshyn, R.W. Storm, "Tracking multiple independent targets: evidence for a parallel tracking mechanism", *Spatial Vision* 3 (1988): 179–197.

Index

Abuse of Cloud Services 21
Access control 28, 38, 41, 71, 90, 148
Access Control Mechanism 90
Access to service 16, 72
Accountability 40, 42, 67
Account Control 20
Account or service hijacking 21
Ad hoc On-Demand Distance Vector Backup Routing (AODV BR) 112
Ad hoc On-Demand Multipath Distance Vector (AOMDV) Protocol 111
Advanced persistent threats 21, 87
Agility 16
Agriculture 83, 84, 85, 133
Alarming/alert engine 26
Anomaly-Based Detection 92
Anonymity 39, 42, 48
AODV nth BR Protocol (Energy-Efficient Routing Technique) 113
Application integrity 19
Application Layer 81, 82, 90
Application Unit (AU) 36, 37
Artificial Intelligence Techniques in Routing Protocol 116
Artificial neural network's structure 123
Asymmetrical network concentration 39
Attacks in green computing 19
Attacks on Cloud Computing 19, 20
Attack simulator 149
Attribute-based encryption framework 153
Authentication 45, 92
Authorization 40, 70, 92
Auto Power Cut 96
Automatic Doors and Windows 84
Automatic Transfer Switch (ATS) 65
Availability 16, 41, 92

Bandwidth Usage 16
Biometrics 73
Blackhole 41, 43
Blockchain 27, 82, 83
Botnets 87
Broadcast tempering 43
Brute force 21, 43
Business Analytics 14
Business Intelligence 14
Business layer 81, 82

Calibrated Cold Vector (CVC) 57
Carbon Emission 4, 5, 18, 66

Carbon footprint 5, 7, 18, 19, 68
Centralized Platforms 27
Cloud Computing 2, 10, 13
Cloud Computing Architecture 11
Cloud Computing Services 12
Cloud Data Center 15
Cloud Layer 11, 14
Cloud Malware Injection attack 21
Cloud solutions providers 23
Cloud system attacks 21
Cluster-based Intrusion Detection 90
Cognizance layer 82
Collision avoidance 35
Communication layer 82
Competence business layer 82
Computing Capacity 16
Conditional privacy 42
Confidentiality 23, 39, 43, 93, 144
Control Identity 28
Controllers 14
Cooling 57
Cooling Units 55
Cooperative communication 46
Critical Load System transfer 65
Cross-site scripting 6, 20
Cryptanalysis attack 87
Cryptographic algorithms 21, 44
Cryptography based authentication schemes 47
Cyber-attacks 6, 20, 88

DAC 153
Dark data centers 69
Data Analysis 13, 16, 83
Data Sharing 164
Data centers 2, 4
Data center Infrastructure Organization Software 62
Data consistency 40
Data control 20
Data Processing layer 14
Decision Tree 122
DELTA framework 142
Deep Learning 6, 27
Demand-compatible self-service 13
Denial of Services 6, 20, 21, 87
Detect Intrusions 24
Digital technology 4
Distributed Denial of Service (DDoS) 86
Distribution of Content 16
DoS 43
Dynamic Source Routing Protocols 108

169

Eavesdropping 43
Edge Computing 10, 13
Edge Layer 11, 14
Efficiency 42
Elliptic Curve Digital Signature Algorithm (ECDSA) 46
Error detection 42
Employee Verification 73
Encryption Attacks 87
End-users 23
Energy conservation 26, 62, 141
Energy Star 2, 5
Enhanced trust-based protocols 27
Entity authentication 41
Environmental impact 39
EPA Green Energy Partner 68
E-waste recycling 2

False attribute possession 43
Faulty System Integration 147
Fault tolerance 45
FleXam 153
FLOVER 153
FlowChecker 153
Focus Layer 82
Fog Computing 10, 13
Fog Layer 11, 14
Fog Sites 15
FortNOX 153
Fostering security 27
FRESCO 153
FTP 144

Gas Sensors 84
Geo-appropriation 16
GLBA (Gramm-Leach Bliley Act) 66
GPS spoofing 43
Green Cloud Computing 17, 19, 27, 127
Green Cloud Security 25
Green Computing 1, 4, 17, 25, 27, 121
Green ICT 1, 5, 121
Green IT 1, 67, 130, 131
Green RFID technology 128
Greyhole 43

Hash-based Encryption 90
Hash functions 46
Health Insurance Liability and Supply Act 66
HiFIND 153
High mobility 39
HomeTec 95
Host Intrusion Detection System 91
Hybrid Intrusion Detection System 92

Identity and data theft 87
Identity-Based Cryptography Protocol 153
Identity management 19
Identity Management Framework 90
IEEE 802.11p 36, 37
Industrial Automation 83, 132
Integrity 40, 93
Information Center Data Committee 68
Information disclosure 43
Information gathering 43
Infrared Sensors 84
Infrastructure-as-a-Service (IaaS) 4, 12
Infrastructure layer 82
Insider attacks 21
Instance-Based Learning 123
Integrated Data Center 58
Intelligent Risk Assessment Based on Neural Networks 153
Intrusion Detection 79
IoT applications 16, 83, 84, 130
IoT architectures 80, 86
IoT based Industrialization 85
IoT Devices 14, 16, 81, 83, 86
IoT security attacks 87
IoT security mechanisms 90
IoT Sensors 15, 83
IR Camera 84

Jamming 23, 43

Key/certificate replication 43
Key Management Center (KMC) 49
K-Nearest Neighbour 123

Latency 16
Liability detection 42
Light Weight Cryptography 90
Location-independent resource management 13
Low slowness 16

Machine Learning 5, 6, 56, 68, 92
Machine learning-based GREEN ICT 129
Magnetic Door Sensors 84
Malicious Insiders 20
Malicious Node Injection 87
Malicious Script 87
Malicious use of machine learning 131
Malware 20, 43
Man-in-the-cloud attacks 21
Man-in-the-middle attack 43, 87
Management Console Security 20
Masquerade 43
Message Authentication Codes (MACs) 46
Message confidentiality 40
Message tempering 43

Index

Microprocessor 10
Microsoft under water Data centers 74
Minimum disclosure 41
MITM attack 87
Mobile Ad hoc Networks (MANET) 34
Mobility model 46
Monitoring 26
Monocrystalline solar panels (MonoSI) 103
MPLS 144
Multi-Cloud Security 27
Multi-Factor Authentication 73
Multitenancy Issues 20
MySpace data centers 19

Naïve Bayes 122
Nanotechnology 3
Network Attacks 87
Network Connectivity 16
Network Fortification 23
Network Heterogeneity 39
Network Intrusion Detection System 91
Network layer 81
Network Security 45
Network security and security framework 142
Network Traffic 16
Neural Network Learning 123
NICE 153
NIDS architecture 154
Nmap 144
Node tempering 87
Non-repudiation 40

On Board Unit (OBU) 36
OpenSAFE 153
OpenSSL 144
Open Web Application Security Project (OWASP) 6
Operational Sustainability 66
Optimization Techniques 26

Pay-per-use facility 13
Perception layer 81
Phishing 6, 87
Physical Attacks 87
PKI Protocols 90
Platform-as-a-Service (PaaS) 4, 12
Polycrystalline solar panels (Poly-SI) 103
Position faking 43
Power Distribution Unit (PDU) 55
Power Usage Effectiveness 26
Power washing data center 75
Preference-based protection 90
Privacy 41, 92
Proactive or Table-Driven Routing Protocols 106
Processing layer 81

Proof of Concept 74
Protocol-Based Intrusion Detection System 91
Public safety 35

Quality of service (QoS) 13

Rack Volume 57
Ransomware 87
Rapid elasticity 13
RE100 68
REBA 68
Reactive or On-Demand Routing Protocols 107
Real time assurances 42
Reinforcement Learning 122
Reliability 16
Relieving Plans 73
Remote recording 87
Remote working 2
Replay 43
Reputation-based Mechanism 90
Revocability 42
RFID Asset Management 73
RFID Cloning 87
RFID Spoofing 87
RFID Unauthorized access 87
Road Side Unit (RSU) 36
Robustness 41, 45
ROI 39
Route diversion 36
Routers 55
Routing attack 87
Routing protocol 39, 45
RSH 144

SAS 70 Type I or II 66
Scalability 41, 45
Secure NBI for SDN applications with NTRU 153
Security Access Protocol 72
Security and Management System (SIEM) 70
Security Threats in IOT 86
SDN risk assessment 147, 150
SDN risk assessment framework 150
SECO 153
Securing distributed SDN controller 153
Sensing Layer 82
Sensors 14, 34, 70, 80
Servers 55
Serverless computers 4
Service layer 82
Service neutralization 145
Side Channel attacks 21
Sign extension 35
Signature based authentication schemes 47
Signature-Based Intrusion Detection 92
Silicon transistors 3

SIMPLE-fying Middlebox Policy Enforcement 153
Simulation of Urban Mobility (SUMO) 48
Sinkhole Attack 87
Site switchgear 65
Smart agriculture 133
Smart Cities 80, 83, 121
Smart Grid 133
Smart LEDs 96
Smart home 83, 88, 93, 95, 99, 102, 132
Smart Home Applications 83
Smart Shopping Trolleys 165
Smart Shower 96
Smart Solar Pane 96
Smart urban communities 133
Smoke Sensors 84
Sniffing 23
SN-SECA 153
Social engineering 87
Software-as-a-Service (SaaS) 4, 12
Software Attacks 87
Software Defined Network (SDN) 142
SOX (Sarbanes Oxley) (2002) 66
Spamming black hole attack 43
Specter and Meltdown 21
Split Multipath Routing (SMR) Protocol 109
Spyware 87
SSL 144
Storage 55
Storage Capacity 16
Storage Constraints 41
STRIDE 153
Structured Query Language (SQL) injection 6
Supervised Learning 121
Support Vector Machine 122
Sybil 43
Synchronization token 21
System Complexity 147

Thermal Cooler 96
Threat at controller 150
Threats at end-host 150

Threats at switch 150
Threats in green computing 19
Three-Layer IOT Architecture 79
Tradeoff 39
Traffic analysis 43, 87
Traffic information 36
Transport layer 81
Trusted Authority (TA) 37
Tunneling 43

Ubiquitous network access 13
Uninterruptable power 55
Unlinkabilty 41
Unsupervised Learning 122
Unused Servers 69

VANETs layered architecture 37
Vehicular Ad hoc Networks (VANETs) 33, 34
Vehicle analytic and maintenance 35
Verification based authentication schemes 47
Video Surveillance 72
Virtual data centers 69
Virtualization 2, 13
VM framework 20
Virus 87
VPN 144

WAVE MAC layer 38
WAVE Management Entity (WME) 38
WAVE Security Service Entity 38
WAVE Short Message Protocol (WSMP) 38
Wearable and Healthcare 83
Weather information 36
Web applications 6
Web service providers 23
Wireless ad hoc networks (WANETs) 34, 105
Wireless Mesh Networks (WMN) 34
Wireless Personal Area Network (WPAN) 34
Wireless Sensors Networks (WSN) 34
Wormhole 43
Worms 87
Wrapping attacks 21